Bergen
Turku
St. Petersburg
Oslo
Stockholm
Copenhagen
Dublin
Hamburg
Tilbury
Amsterdam
Bordeaux
Nice
Genoa
Venice
Marseille
Livorno
Dubrovnik
Barcelona
Portofino
Civitavecchia
Istanbul
Lisbon
Palma
Naples
Piraeus
Izmir
Tianjin
Tunis
Santorini
Yokohama
Tangier
Mykonos
Rhodes
Crete
Casablanca
Ashdod
Madeira
Shanghai
Canary Islands
Taipei
Hong Kong
Bombay
PACIFIC
OCEAN
Rangoon
Manila
ARABIAN
SEA
Bangkok
Saigon
Phuket
Kuala Lumpur
Singapore
Mombasa
Mahé
INDIAN
OCEAN
Bali
Brisbane
Durban
Perth
Cape Town
Sydney
Melbourne
Hobart

THE WORLD'S 100 MOST EXCITING PORTS OF CALL
SELECTED BY THE WORLD'S LEADING TRAVEL WRITERS

V·O·Y·A·G·E·S

The Romance of Cruising

A TEHABI BOOK

THE WORLD'S 100 MOST EXCITING PORTS OF CALL

SELECTED BY THE WORLD'S LEADING TRAVEL WRITERS

A·G·E·S

The Romance of Cruising

PHOTOGRAPHS BY HARVEY LLOYD

A TEHABI BOOK

ACKNOWLEDGMENTS

My sincere and heartfelt thanks to Ross Eberman for his untiring efforts to make this book one we can all be proud of; to Tom Lewis whose splendid design of *Voyages* exceeded my expectations; to the lovely artist Shirlee Price who worked side by side with me on countless missions around the world. In addition I would like to thank my friend Spencer Frazier for making many of these images possible and my son, Andrei Lloyd, for encouragement. Thanks as well to Chris Capen and all the fine craftsmen and women at Tehabi Books; my many friends in the cruise industry; the generous cruise ships' captains and crews; the daring pilots who flew me around these ships and above landscapes and seascapes; and everyone whose blood tingles at the sound of a ship's horn and who find joy and adventure at ports of call around the world. —HARVEY LLOYD, NEW YORK CITY

TEHABI BOOKS

A DK PUBLISHING BOOK
www.dk.com

Alrica Goldstein, Project Editor
Dirk Kaufman, Art Director
LaVonne Carlson, Editorial Director

First Edition
10 9 8 7 6 5 4 3 2 1
Published in the United States by
DK Publishing, Inc.
95 Madison Avenue
New York, New York 10016

First published in Great Britain in 1999 by
Dorling Kindersley Limited, 9 Henrietta Street, London WC2E 8PS
A CIP catalog record of this book is available from the British Library.
ISBN 0-7513-0807-2

Library of Congress Cataloging-in-Publication Data
Lloyd, Harvey.
Voyages : the romance of cruising / photographs by Harvey Lloyd.
— 1st American ed.
p. cm.
ISBN 0-7894-4617-0 (alk. paper)
1. Ocean travel. I. Title.
G550.L57 1999
910.4'5—dc21 99-15272
CIP

Printed and bound in Korea through Dai Nippon Printing Company

Voyages was conceived and produced by Tehabi Books. *Tehabi*—symbolizing the spirit of teamwork—derives its name from the Hopi Indian tribe of the southwestern United States. As an award-winning book producer, Tehabi works with national and international publishers, corporations, institutions, and non-profit groups to identify, develop, and implement comprehensive publishing programs. Tehabi Books is located in Del Mar, California. www.tehabi.com

Chris Capen, President
Tom Lewis, Editorial and Design Director
Sharon Lewis, Controller
Nancy Cash, Managing Editor
Andy Lewis, Senior Art Director
Sarah Morgans, Associate Editor
Mo Latimer, Editorial Assistant
Maria Medina, Administrative Assistant
Steve Lux, Art Director
Kevin Giontzeneli, Production Artist
Sam Lewis, Webmaster
Ross Eberman, Director of Custom Publishing
Tim Connolly, Sales and Marketing Manager
Eric Smith, Marketing Assistant
Tiffany Smith, Executive Assistant

Jay Clarke, Consulting Editor
Gail Fink, Copy Editor
Kathi George, Proofreader
Ken DellaPenta, Indexer
Laura Georgakakos, Copywriter
Joe Yogerst, Copywriter

Tehabi Books offers special discounts for bulk purchases for sales promotions or premiums. Specific, large quantity needs can be met with special editions, including personalized covers, excerpts of existing materials, and corporate imprints. For more information, contact Tehabi Books, 1201 Camino Del Mar, Suite 100, Del Mar, CA 92014, (800) 243-7259

Pages 2–3: **S***eabourn Legend sails beneath the Golden Gate Bridge in San Francisco.*

Pages 4–5: **T***he* Windstar *cruises off lush St. Kitts on the way to Nevis.*

Opposite: **P***assengers line the decks to get a keener view of Norway's Troll Fjord.*

Pages 8–9: **A** *Japanese fireboat renders a dramatic sendoff to* Royal Viking Sun *(now* Seabourn Sun*) departing Osaka.*

Pages 10–11: **P***assengers on Holland America's* Veendam *take in some of the glories of the past as they cruise through historic ports in the Caribbean.*

Pages 12–13: **T***he* CostaRomantica *heads out into the open sea*

Pages 14–15: **A** *chromatic sunset over the Caribbean suffuses the "Fun Ship"* Ecstacy*.*

Holland America Line

COSTA CLASSICA

ECSTASY

Cunard
White Star

Contents

The maiden voyage of Cunard White Star Line's Queen Mary *in 1936 is commemorated in a vintage poster.*

BY JAY CLARKE

BON VOYAGE

RATING THE WORLD'S BEST

EVERY TIME I SAIL into port on a cruise ship, a surge of anticipation builds inside me. I know I'm about to embark on another extraordinary adventure. I revel at the thought of the new places and new faces I'll encounter. I look forward to immersing myself in a different culture, eating out-of-the-ordinary foods, shopping for exotic goods, and seeing sights and scenes totally unlike those at home. The new port beckons, and where it will lead me I never know until I step ashore and venture beyond the pier.

As a travel editor, I've done that a lot. And while I haven't visited every port around the world, I have swapped travel stories and traded tips with hundreds of other travel writers about favorite places and favorite experiences. That kind of tale-telling never fails to set off my adrenaline—suddenly I'm chafing to get going on another voyage.

That's why I'm so excited about this book. With its combination of extraordinary photos and inside tips from experienced travelers, *Voyages* is the perfect companion for those who enjoy travel, be they experienced cruisers, prospective sailors, or armchair travelers. Not a guidebook, it is a lavish look at the best that worldwide cruising has to offer, illuminated with revealing photographs and insider details from the real travel buffs—the people who make their living visiting and writing about destinations all over the world.

WE EXPLORE THE MYSTERIOUS CASBAH OF CASABLANCA, STEP INTO THE GOLD RUSH SALOONS OF SKAGWAY IN ALASKA, WINDOW-SHOP ON THE STORE-LINED STREETS OF ST. THOMAS . . .

In his introduction, John Maxtone-Graham, the eminent author of *The Only Way to Cross* and many other books and articles about passenger ships, sets the tone with tales of his best-remembered ports. Bill Miller, long a respected delineator of the seafaring experience, goes back to the turn-of-the-century beginnings of cruising, then brings us to the booming modern era.

Dividing the world into eleven regions, expert essayists then paint a vivid picture of each destination as well as the cruising experience. We follow them as they venture beyond the water's edge into the streets and surroundings of a hundred ports

Passengers bask in the sunshine over the Indian Ocean, enjoying a popular pastime on many cruises.

of call. We explore with them the mysterious Casbah of Casablanca, stroll on Barcelona's romantic tree-lined Ramblas, step into the gold rush saloons of Skagway in Alaska, window-shop on the store-lined streets of St. Thomas, and pick our way through the majestic ruins of Ephesus in Turkey.

THE TEN MOST EXCITING
PORTS OF CALL

1. *New York*
2. *Hong Kong*
3. *San Francisco*
4. *Venice*
5. *New Orleans*
6. *Nice, Cannes, Monaco*
7. *Istanbul*
8. *Bangkok*
9. *Buenos Aires*
10. *Tilbury (London)*

In their quest for exciting sights, cultures, and peoples, they rated ports with such unfamiliar names as Civitavecchia and Tianjin among the world's best, as well as places known to us in other contexts: Antigua, the island where Britain's Admiral Horatio Nelson made his Caribbean base; Cancún, the sun-blessed gateway to the fascinating Maya ruins of Mexico's Yucatan peninsula; Honolulu's Pearl Harbor, the scene of the attack that brought the United States into World War II; and Rio de Janeiro, the site of what many consider the world's most beautiful harbor. And then there are some extremely familiar port cities that were not rated among the one hundred best. Popular destinations like Juneau, Auckland, Nassau, Grand Cayman, and Cartagena were very close to the cut-off point, but their overall averages barely missed the mark. That's the fun of this sort of undertaking: the unexpectedness of some of the results.

New York was selected as the most exciting port in the world.

All told, through the eyes of these writers, we visit a hundred places where today's cruise ships call. These featured ports are not random selections. More than 100 travel writers responded to our survey, which asked them to name the world's best port cities in which to experience adventure, entertainment, cuisine, romance, and shopping. You'll find their ratings—from one to five in each category—with each of the one hundred profiled ports.

New Orleans was the port chosen for the best food in the world.

The survey results were taken a bit further than just determining the world's one hundred best ports; the top ten ports for each of the five categories were calculated. According to the travel writers, we should do our shopping in Hong Kong, feast in New Orleans, experience the entertainment offerings of New York, and take our lovers to Venice. And for adventure, we should voyage to a port that topped that category's list but did not make the overall top one hundred: the Galapagos Islands. Exploring these islands is not just high adventure, it's an unforgettable experience. Imagine coming face to face with giant tortoises, marine iguanas, and even penquins!

THE TOP TEN PORTS FOR
CUISINE

1. *New Orleans*
2. *New York*
3. *San Francisco*
4. *Hong Kong*
5. *Nice, Cannes, Monaco*
6. *Bordeaux*
7. *Venice*
8. *Singapore*
9. *Livorno (Florence)*
10. *Bangkok*

Probably the most intriguing information to come out of this survey is that of the top ten most exciting ports of call in the world.

THE TOP TEN PORTS FOR
ADVENTURE

1. *Galapagos Islands*
2. *Istanbul*
3. *New York*
4. *San Francisco*
5. *Mahé, Seychelles Islands*
6. *Sydney*
7. *Hong Kong*
8. *Venice*
9. *Tilbury (London)*
10. *Nice, Cannes, Monaco*

The Galapagos Islands are rated tops for adventure.

A *day sailor and Cunard's* Queen Elizabeth 2 *cross paths off* St. Thomas.

The port city that received the highest overall rating is New York. The Big Apple almost has it all: It made the top ten lists in every category except Romance.

The stunning photographs that give the book such brilliant dimension were taken by the internationally renowned Harvey Lloyd, who is generally regarded as the pre-eminent cruise photographer in the world. Lloyd has traveled more than a million and a half miles on assignments, and his images capture the essence of the ports with dramatic effect, sometimes with views of the urban landscape, sometimes closing in with more intimate portrayals of people, customs, cuisine, and culture. It's an extraordinary visual adventure.

THE TOP TEN PORTS FOR **ENTERTAINMENT**

1. *New York*
2. *New Orleans*
3. *Los Angeles*
4. *San Francisco*
5. *Tilbury (London)*
6. *Buenos Aires*
7. *Rio de Janeiro*
8. *Sydney*
9. *Miami*
10. *Hong Kong*

Venice is the most romantic of ports.

In the century since cruising for pleasure got its start, the world has undergone radical transformation. So, too, has cruising. Sea travel is no longer the ordeal it was a century ago; today's liners are luxurious cocoons sailing into every body of water on earth. Ports of call have changed as well. More than ever, they are conduits to peoples and cultures we hunger to understand. Once-remote destinations that were unattainable dreams for most people now are easily visited. And whether we are dreaming about our next—or first—cruise, this book lets us explore the globe as though we were on an around-the-world journey.

Bon voyage! ⚓

THE TOP TEN PORTS FOR **ROMANCE**

1. *Venice*
2. *Mahé, Seychelles Islands*
3. *Nice, Cannes, Monaco*
4. *Bali*
5. *Portofino*
6. *San Francisco*
7. *Santorini*
8. *Mykonos*
9. *Bermuda*
10. *Quebec City*

THE TOP TEN PORTS FOR **SHOPPING**

1. *Hong Kong*
2. *New York*
3. *Bangkok*
4. *Istanbul*
5. *Singapore*
6. *St. Thomas*
7. *Venice*
8. *Bali*
9. *Tilbury (London)*
10. *Livorno (Florence)*

The Caribbean continues to be a favorite cruise ship destination.

Hong Kong is still considered the best place for shopping.

B Y J O H N M A X T O N E - G R A H A M

V·O·Y·A·G·E·S

ABOARD AND ABROAD

The sea is at its best at London, near midnight, when you are within the arms of a capacious chair, before a glowing fire, selecting phases of the voyage you will never make.

—H. M. TOMLINSON, *THE SEA AND THE JUNGLE*, 1912

ONE TRIBUTARY OF THE mighty Rhine flows through the Swiss city of Basel as a burbling river, no more than fifty feet wide, slipping in sepia-colored sheets across a pebbled bottom. It can only be crossed aboard a minuscule but efficient ferry, for pedestrians only and powered by the current. Attached ten feet up the trunks of two opposing trees, a rigid wire spans the flood. The ferry, no more than a stout skiff, has its painter looped around that wire with a slip noose. One embarks and the ferry's master casts off and puts his rudder hard over; the current, impinging three-quarters on the bow, thrusts the ferry obliquely but steadily to the farther shore.

Since the ride lasts moments and costs a pittance, my children and I were indulged by the ferryman with several consecutive crossings. Let me confess, ferries of any kind—whether crossing rivers, channels, or oceans—have always intrigued me. Another that we used to enjoy was a chain ferry connecting the banks of the river Itchen in Southampton on England's south coast. Far larger than the Basel skiff, this one boasted a noisy diesel engine that methodically devoured a length of tethered chain drawn up from the river bottom, pulling along the ferry loaded with cars, trucks, and pedestrians, including this ferry fanatic and his family.

WHEN ASKED BY INTERVIEWING JOURNALISTS MY FAVORITE SHIP, I INVARIABLY REPLY, "THE ONE I'M ON"; SO, TOO, IF ASKED MY FAVORITE PORT, I FEEL COMPELLED TO SUGGEST, "THE ONE I'M IN."

That chain ferry is no more, having been supplanted by a soaring bridge along which the same traffic that once queued patiently now zips impatiently overhead. The superannuated chain is doubtless rusting away in the Itchen's mud.

But consider this allegorical parallel: whereas those Southampton motorists have been denied their sea-level crossing in favor of a prosaic, elevated route, contemporary travelers

Captain Ola Harshein of the Royal Viking Sun *joins passengers to wave a greeting off Geiranger Fjord in Norway.*

bound for shores are still afforded a choice: either passage by sea or the aerial alternative. Every one of the ports listed within this volume can be reached more quickly and less expensively by plane, but I prefer without question the old-fashioned, languid option of a seaborne approach. Fly if you must, but be warned that your swifter journey will be devoid of the epic satisfaction of arrival by water.

An almost cliché Victorian cure-all was the tonic, as it was known, of a sea voyage. In one sense, it must have been an ambiguous remedy for Britons because for them, ocean passage in every direction involved fractious waters, whether east across the North Sea, south around France through the dread Bay of Biscay, or west for three thousand miles of uncomfortable steaming to America. By contrast, today's shipboard affords even the rankest landlubber the privilege of embarking for distant lands in stabilized comfort, cossetted by air-conditioning, stewards, and Dramamine all the way.

In truth, ferry fixations are fueled by the lure of destination. By good fortune, my life for the past three decades has remained inextricably entwined with the attainment of abroad by sea. The going is easy, for today's cruise ships have never been more numerous, nor as large, comfortable, and sometimes luxurious. What enchants their occupants are the time-honored, satisfying rituals of shipboard. And though cruise lines insist that their vessels are destinations in and of themselves, we know better. Cruise ships are far more: a seductive—nay, intoxicating!—means of achieving exotica.

Some of the world's most stunning sights are only visible from the deck of a ship.

What compels us to embark?

Whence comes the pull that seduces us abroad? There is not one answer but many. For starters, sated by our increasingly homogenized, day-to-day diet, all of us crave periodic infusions of visual, aural, and tactile spice. Within the familiar cocoon enclosing us, we are too often deprived of the sights, sounds, and smells so typical of remote landfalls. Though much of our world has been comprehensively shrunk and documented by a miraculous web of communication, no brochure, film, video, or internet browse can replicate the impact of firsthand immersion within alien climes. Rather than broadening the mind, travel broadens the senses, imparting novel inputs to jaded palates and rewarding us with exposure to contrasting lives and locales.

Then, too, surely we are drawn to foreign parts by the promise of adventure, where strange tongues and strange ways will test our traveling mettle. It's not all paradise, nor should it be. No one who has toured the crumbling municipal infrastructure of Vladivostok, been hounded by Bombay's desperate beggars, or wandered the length of Cairo's teeming souk can reembark at day's end less than profoundly distressed or moved. Exotic ports are more than picture postcards; many come with a warts-and-all pungency that we must absorb and accept as inevitably part of the day's outing.

After more than half a century of shipboard, I would be at a loss to name my favorite port. How very sensible of the publishers of this volume to incorporate a hundred. When asked by interviewing journalists my favorite ship, I invariably reply, "The one I'm on"; so, too, if asked my favorite port, I feel compelled to suggest, "The one I'm in." But several among dozens remain memorably within my mind.

I shall never forget Shanghai over the winter of 1981, achieved aboard *Rotterdam*, where the muddy outflux from the Yangtze and Whangpoo rivers marbleizes the offshore waters of the East China Sea. Inside the great harbor, the gritty air hung heavy with a sulfurous and, since I was raised in coal-fired London, nostalgic miasma from thousands of the port's funnels and chimneys.

Ashore in those Mao days, the streets and the populace were utterly colorless, awash with quilted, dun-colored civilian dress, the soldiers' red collar insignia the only flash of color anywhere. Such was the unisexual predilection for trousers that my wife Mary's stockinged legs showing below the hem of her tartan skirt drew crowds wherever we went. Sheltering from the cold in a waterfront teahouse, we shared a table with a gigantic private soldier who, with enviable dexterity, devoured two trembling slices of lemon meringue pie with chopsticks.

We were aboard *Rotterdam* another year, alongside in Colombo, entry port into the recently troubled island of Sri Lanka. I always love browsing through those instant, makeshift markets that, in the East,

Right: Under a full moon, Seabourn Pride *cruises off St. Thomas in the U.S. Virgin Islands.*

spring up in the humid shade beneath cruise ship bows. One persistent stall holder offered for sale an entrancing ivory elephant, caparisoned with miniature Candian finery. Although it had taken my fancy, I am a hopeless bargainer. But as the afternoon sun swung 'round and the ship's departure approached, passing time worked the haggling miracle that I had been helpless to achieve: the price dropped precipitously.

When we sailed out into one of those extravagant Indian Ocean sunsets, the handsome ivory stood on my porthole ledge. And back on the pier, the stall-holder had doubtless pocketed exactly what he had originally anticipated as his elephantine due.

The elegant CostaRomantica *shines against the whimsical colors of Venice.*

I first visited Venice by ship during *Grand Princess*'s maiden voyage. After a marvelous lunch overlooking the Grand Canal from the terrace of the Gritti Palace Hotel, Mary and I joined the tourist throngs drifting across bridge and square toward the Piazza San Marco. Although it was Sunday, most of the shops were open and, not surprisingly, a roped red, blue, and gold necklace literally shouted to us both from a jeweler's window. Within moments—no haggling here—the necklace was firmly in place around Mary's neck and we celebrated within teeming, pigeon-swarming St. Mark's with an overpriced cappuccino next to the orchestra at Quadri's as the lights came on.

Not all my favorite ports are merely acquisitive. One of the world's port-richest cruises is aboard a companionable coastal steamer sailing up Norway's coast to the North Cape and back. Departure port Bergen makes for a splendid municipal overture. At the end of the Vaargen, a long sea arm penetrating the city, is a market selling flowers and fish. We carried a plump sea trout to a restaurant behind the Bryggen, that historic row of Hanseatic houses, where the Greek chef grilled it to perfection.

Passage north features Norway's incomparable shore scenery for all twenty-four hours of the day; forested southern regions ultimately segue into bald mountain slopes dotted sparsely with hauntingly isolated farms or fishermen's cottages. The vessel calls at three or four ports each day; though not all are memorable, the cumulative sea/land interface is beguiling.

The country's bleak, northernmost eminence had changed drastically since my last visit. The North Cape has been effectively malled, transformed into an extensive eating/drinking/shopping emporium gouged vertically down into the Nordic granite; there is even a luxurious cinema auditorium at the bottom. In one restaurant, a huge plateglass window penetrates the north-facing cliff, offering a protected vantage point toward (if not in sight of) the boundless arctic ice and its elusive pole. Picture window or no, the North Cape was shrouded in fog that day and we saw only opaque white.

We docked once at Hiroshima, walking inland beneath an autumn sun to the nearest tram stop that would carry us to the war memorial, site of the first but not quite the last (Nagasaki followed) atomic weapon detonated over an essentially civilian target. It was not difficult determining fares or destination, thanks to an extremely cordial and punctilious driver. Our gleaming green conveyance rumbled pleasantly and purposefully through Hiroshima's bustling streets en route to a mere block away from the great museum/memorial within a park.

Sharing our visit were a few western tourists and dozens of uniformed school children who added their tributes to a low wall literally upholstered with thousands of colorful prayer offerings long in place. An obliging Japanese offered to take our picture together before the stark, devastated skeleton of Hiroshima's town hall, ground zero for the nuclear carnage that had incinerated thousands of his countrymen; it was he who insisted that we smile.

There are few entries into fabled ports more imposing than the Sydney heads, towering cliffs that, north and south, serve as sentinels giving access into the commodious harbor of Australia's principal city. Greenery-topped cliffs relent into Sydney proper, and the stunning, white-petaled roof of

Seabourn Goddess I *cruises off an island in the Caribbean.*

The thrills experienced while cruising aboard a traditional sailing ship such as Windstar *are never to be forgotten.*

A *rainbow breaks over Charlotte Amalie on St. Thomas in the U.S. Virgin Islands.*

the opera house—unquestionably the world's most recognizable cultural icon—lay across the water from our pier. It happened to be Anzac Day, Australia's Armed Forces and Memorial Day combined, as well as the only day all year when the celebrated army gambling game—Two Up—can legally be played. Bemedaled veterans were everywhere and the pubs were packed to overflowing, spilling hordes of tankard-laden clients out onto the pavement. Security officers kept the peace while guiding nondrinking pedestrians past their domain via the gutter.

There was time both to buy an opal and enjoy a string quartet playing Bach in a pedestrian square. That evening, we washed down two dozen of Sydney's sweet rock oysters with a flagon of Guiness at one of the restaurants lining the illuminated quay, along which our *Island Princess* shared pride of place with a flawless facsimile of HMS *Bounty*.

The incredible splendor of Glacier Bay in Alaska encircles passengers aboard Holland America's Statendam.

Glancing back over this roster of ports, it occurs to me that an inordinate amount of my time ashore seems to revolve around memorable purchases or meals. Of course, one should never neglect cultural riches at their expense and, in the Mediterranean particularly, an almost infinite selection of cathedrals, museums, and palaces remains etched firmly within my memory. But whereas they are, essentially, immutable fixtures that will always be in place, my serendipitous quests through narrow shopping streets or restaurant tables remain fragile, impermanent outings, prone to vanish within the roseate afterglow of the completed cruise. Moreover, whatever one purchases in any port of call remains forever attached in the mind to that specific, exhilarating day.

My final favorite is, inevitably, the most acquisitive of all: Hong Kong, irresistible, neon-fired China. Although the bargains of yesteryear are harder to come by, it is the very vigor of the place, from dynamic skyline to beetling harbor traffic, that seems to resonate throughout the vessel as we tie up alongside the Ocean Terminal. There are shops,

Masks are an important part of the ritualistic dance and folk drama of Bali.

literally, at the foot of the gangway and there are more shops, shopping centers, and markets everywhere; "retail therapy," deadpanned one fellow passenger. They used to say that when world cruise ships tied up for their requisite three days (long enough for completion of a bespoke suit), shopkeepers throughout the colony would rake in over a million dollars. And Orson Welles once chided his shipbound fellow passengers, suggesting that if they weren't shopping ashore, they were wasting money.

Whether the clanging trams of Nathan Road or the glittering riches of the Chinese Arts and Crafts Store; whether the string orchestra accompanying tea within the Peninsula Hotel's venerable lobby or the cool elegance of lunch atop the Mandarin or the vast Ocean Palace where those seductive little dim sum carts wend their persuasive way among hundreds of tables, augmenting and reaugmenting what started as a light lunch—fragrant, hustling Hong Kong enchants. And stitching the harbor waters separating mainland Kowloon from Hong Kong Island are the doughty, green-and-white vessels of the Star Ferry that, again for a pittance, carry commuters and tourists back and forth all day and much of the night.

We began this smorgasbord of ports with a ferry and with another it should end, symbolic of that leap of faith and pleasure that anticipates ports to come. Countless times have I stood on the bridge of a ship before dawn and spied, in the distance, land's faint outline materializing in the half-light. There comes a moment when those distant twinkling lights scattered across island slopes mirror precisely the glowing electronic pinpoints illuminating the bridge's navigation and control dials. That momentary, privately observed juxtaposition always delights me. Surely it can be no accident that *aboard* is an anagram of *abroad*, firm lexical endorsement of those beloved ships that convey us so splendidly to ports around the world. ⚓

Skirting the islands on Alaska's Inland Passage on the way to Sitka affords spectacular views of natural beauty.

Pisa is a city filled with romance, history, and art, but is best known for its famous leaning tower.

Right: **W**ith twinkling lights in its masts, Windstar prepares to sail on a Caribbean cruise at sunset.

Previous page: **T**he Royal Viking Sun *at Venice basks in the golden glow of dawn over the Adriatic.*

A *lounge chair beckons poolside on the upper deck of the* CostaRomantica.

WHITE STAR LINE
THE WORLD'S LARGEST LINER
SOUTHAMPTON
CHERBOURG
NEW YORK

BY WILLIAM MILLER

C·R·U·I·S·I·N·G

THE ROMANTIC HISTORY OF VOYAGING

MILLIONS AGREE: THERE IS no vacation experience quite like a cruise. From the very first moment you step aboard, you are transported to a different world. And an adventure begins! A sense of romance, of a different lifestyle and of a different time, takes hold. Actually, time takes on new measurements: the clock becomes almost insignificant while life becomes paced, more casual, certainly more gracious.

Today, cruising is almost the complete mainstay of the ocean liner business. Even the *Queen Elizabeth 2*, which still offers transatlantic crossings, is run more like a cruise ship than one of those traditional steamers of yesteryear. Moving travelers from point A to point B has now become the province of aircraft.

On sailing day aboard current cruise ships, you are welcomed on board by smiling, helpful staff members and taken with anticipation as well as reassurance to your cabin, your home away from home for the duration of the voyage. The sense of security, of belonging, takes hold. You are embraced by the ship, the crew, the officers. It is almost maternal, womblike. All the weight of responsibility is lifted. Someone else is taking charge of your well-being, your security, your happiness. Later, as bags arrive, clothes fill the closets, and personal effects are scattered about, those perceptions of security and belonging are heightened. The ship is your hometown, the cabin your residence. A feeling of proprietorship emerges, as if the whole ship becomes yours.

THE WHISTLE SOUNDS. DEPARTURE IS CLOSE AT HAND AND EXCITEMENT BUILDS. THE EXPERIENCE OF AN OCEAN VOYAGE, NO MATTER HOW BRIEF, IS ABOUT TO BEGIN.

The whistle sounds, signaling that departure is close at hand. Excitement builds while reality—the cares and worries and stresses of life ashore—quickly begins to fade. The experience of an ocean voyage, no matter how brief, is about to begin. When that last rope to the dockside is released and the ship begins to move, you realize you are on a floating city. It is much like the release of a balloon. You are aboard a great moving object; something once solid and inactive now moves. There is a sense of freedom, disconnecting, letting go.

The White Star Line proudly advertised its luxurious Majestic, *the world's largest ocean liner of the 1920s.*

As the ship undocks, often with the help and reassurance of a sturdy little tug, you feel the gentle hum and momentary vibration of the ship's engines. Suddenly, there is great power beneath you, decks below, but in other, more competent hands.

Sailing away is always a thrill. It can actually create a tingle down your spine. As the ship moves along the harbor and heads for the open sea, these are magical moments. Shoreside diminishes, often in a quite romantic form: in a hazy mist or a blazing sunset or a blanket of surreal fog. The wind and the salt spray hit your face. Flags attached to the mast flutter and snap. The ship picks up speed. The adventure has truly begun.

The Royal Viking Sun's Stella Polaris Lounge is an example of the sumptuous luxury passengers enjoy on today's cruise ships.

Later, passengers explore and uncover more of these vast moving complexes. They wander about the decks, uncovering the lounges and restaurants, the gift shops and spa

facilities, the lido areas and those long, orderly corridors, the main reception and the shore excursion office. The sense of "grand hotel" unfolds quickly; everyone seems to be scurrying about. Fellow guests exchange their first nods and smiles, and possibly initial conversations. A friendliness prevails.

Once at sea, shipboard takes on yet a different tone. A relaxed, calmer, even quieter mood often takes hold. There are no worldly intrusions: no ringing phones or harsh news or traffic jams or rushing about to meet deadlines. Quite simply, there are few, if any, pressures. Instead, it all becomes about enjoyment and pleasurable pursuits: leisurely living and fine dining, reading, a nap, enjoying other people, seeing a show, dancing, and simply observing the sea and the sky around you. Sometimes, thoughts become more transcendental, wandering to senses of art and speed and light. And, of course, a sense of deepening romance takes hold. There is a great beauty in the sea and the sky, and sheer wonder in their vastness and qualities of change.

The
BLUE FUNNEL
LINE
TO CANARY Is
SOUTH AFRICA
AUSTRALIA EGYPT
COLOMBO STRAITS
CHINA & JAPAN
KENNETH D SHOESMITH
ALFRED HOLT & CO., India Buildings LIVERPOOL
Telegrams: "ODYSSEY."
Telephone: Bank 300.

And no matter the size of the ship itself, the enormity of our earth becomes evident. The ship is transformed into a little piece of steel by comparison. But in itself, it remains wondrous, a powerful machine moving through the water, sometimes at incredible speed. There is often an awareness of the past, of seafaring through the ages, and of the timeless traditions of the sea.

* * *

The very first cruise is said to have been offered in 1858. It was a sightseeing voyage of sorts to the antiquities of the then-remote, very exotic Mediterranean. A British ship, the *Ceylon*, owned by Peninsula & Oriental Steam Navigation Company Limited (P&O), was used for this voyage. It was all rather primitive and basic, however. It was long before air-conditioned staterooms, lido decks with oval-shaped pools, or even the luxury of comfortable onshore tour buses for ferrying passengers to all those temples and treasures. Indeed, the passengers on that first cruise aboard the *Ceylon* were adventurous and filled with the spirit of exploring. And they were certainly well heeled! Menu items included salt fish and egg sauce, a roasted haunch of mutton, and currant fritters.

Actually, much earlier, in 1815, an advertisement was run in an Edinburgh newspaper calling for "tourists" and proposing a "cruise" to the remote Faeroe Islands and then onward to Iceland. But it was far too adventurous for most would-be travelers of the time and so the voyage never quite set off.

The flower and plant stewardess is but one member of the crew serving and assisting passengers on every cruise.

Following the *Ceylon's* inaugural pleasure jaunt, cruises remained rather intermittent and certainly select. They were daring journeys, still restricted to the more adventurous souls or, as one advertisement called them, "the educational and scientifically oriented." But by the twilight of the Victorian age, in the 1890s, there were increasing numbers of these so-called adventurers who had both the time and money to take one of these "voyages of rest and recreation." One company, Germany's Hamburg-America Line, already one of the world's mightiest and most innovative passenger ship operators, was keen on pursuing this market. Among other factors, they realized their traditional North Atlantic service to New York in the cold, often storm-tossed winter months slumped. There were fewer passengers going from the Old World to the New and consequently there were fewer profits. So they began to dispatch some of the smaller passenger vessels in their fleet on pleasure voyages to the West Indies, the Mediterranean, West Africa, and even far-off Egypt. One cruise, on the eight-thousand-ton *Augusta Victoria*, left Hamburg-America on a two-month journey in January 1891. One of the biggest ships of her day, this liner was equal to five thousand present-day automobiles.

The *Empress of Japan* was the fastest liner on the Pacific in the 1930s.

It was in fact the ingenious Hamburg-America Line that, with considerable foresight and courage, built the world's very first cruise ship. She was the forty-four-hundred-ton *Prinzessin Victoria Luise*, completed in 1900. Although only just over four hundred feet in length, her small size was rather deliberate: she was purposely designed to resemble the royal yachts of Europe. After all, her clientele was none other than the very rich—travelers who could afford to spend weeks, if not months, aboard and who wanted only the very finest food and service. There were 119 staterooms on board that could, altogether, berth four hundred guests. For the first time on any ship of any kind and size, there was not only a complete bedroom and sitting room in each stateroom, but a private bath as well! There was even a

Glorious sunsets are enjoyed amid luxurious comfort.

The Blue Funnel Line promoted exotic voyages from the United Kingdom to the Far East in the 1930s.

North German Lloyd Express North Atlantic steamers set speed records during the late 1920s and early 1930s.

special royal apartment on board, made available, should the desire arise, to the kaiser himself. Sadly, however, life was quite short for the world's first cruise ship. Within six years, in 1906, she went aground on Jamaica and became a complete wreck.

By 1910–11, Hamburg-America Line was even running cruises "around the world," trips that began at New York and then ended one hundred days later, at San Francisco. (Return to New York was made by train.) These immediately proved so popular that two trips had to be run each winter. Fares ranged from $6,000 per person for a suite with a bedroom, sitting area, and private bath to $650 for "gentlemen only" in smaller, formerly third-class quarters down on F deck.

After the First World War, in the 1920s, cruising expanded even further. But it was still for a select following, a very specific clientele, the "millionaire set" as they were called. In that age, long before the rigorous, often harsh economics of passenger ship operations, some of these two hundred "cruisers" might be looked after by as many as eight hundred crew! And now, even the world's largest and most celebrated liners, such as Cunard's *Mauretania*, the fastest vessel afloat, and their *Aquitania*, so splendid that she was dubbed "the ship beautiful," sought off-season employment away from that dreary, often storm-ridden, North Atlantic run. These ships were, if only temporarily, made over into all-first-class "floating clubs" and sent off to the warm waters of the Caribbean or on more exotic Mediterranean itineraries. The *Mauretania*, for example, would sail on an eight-week cruise, departing and then returning to New York, and calling in at such ports as Casablanca, Gibraltar, and Villefranche on the French Riviera. Other legendary Atlantic liners followed in her wake—the impeccably fed *Paris*, the spotlessly maintained *Rotterdam*, and the flawlessly served *Homeric*.

The CostaMarina *carries passengers into another evening of moonlight and music.*

Expectedly, brand-new and increasingly lavish passenger ships soon sailed off on similar sojourns. Cunard's splendid *Franconia*, for example, was commissioned in 1923. Each winter, she headed for the West Indies and South America—to Havana and Trinidad and Rio. Her accommodations were especially fine and featured a two-deck-high smoking room fashioned after El Greco's fifteenth-century residence. Another cruising Cunarder, the *Laconia*, had a smoking room done to resemble an Old English inn; the decor included a red brick fireplace!

It was in the otherwise hard-pressed, depression-era 1930s that cruising first caught the imagination of the general public and expanded greatly. Short voyages became escapes for many, even if only temporarily. There were quick runs out of New York to Bermuda, Nassau, and Havana and—in summer—up to Halifax. But cheapest and therefore most popular of all were the so-called booze cruises, party trips to nowhere. Leaving New York in late afternoon and then heading out past the three-mile territorial limits, foreign-flag cruise ships had bars that could be opened and thereby escape the rigors of America's Prohibition. These cruises were priced from $10 and often were aboard such illustrious Atlantic liners as the *Mauretania*, *Berengaria*, and *Majestic*. And from these cruises, thousands later graduated to longer voyages: four days to Bermuda and back, six days to Nassau, eight days to Havana.

Nothing compares to the peace and escape offered by several days at sea.

Ocean liner historian Everett Viez was aboard the *Aquitania* at this time. "Cunard still ran a top-rate ship, even in the bleak, money-short depression. Service, cuisine, and of course the ship itself were quite exquisite," he recalled. "We were on a four-and-a-half-day, Fourth-of-July, long weekend cruise. We sailed from New York on Friday at midnight, sailed up to Halifax, and returned to New York on Tuesday evening at six. It was $50 per person in the cheapest accommodations. She was such a magnificent ship, but an old-timer by then. There was a working fireplace, using anthracite coal, in the Elizabethan Smoking Room."

Another British company, the Furness-Bermuda Line, also did much to popularize cruising, in particular the trade to Bermuda, in the 1930s. It is a mere forty hours by ship from New York, creating an appealing six-day round-trip voyage with three full days on the island

Overleaf: Passengers enjoy good food and good times in the La Fontaine dining room on board Holland America's Statendam.

Fascination
Carnival

itself. The company added two fine liners, the *Monarch of Bermuda* and then the *Queen of Bermuda*, in 1931–33. Purposely designed to resemble the big transatlantic liners, they were very popular from the start and were often fully booked months in advance. There were two sailings from New York each week, but their Saturday afternoon sailings were so popular with the just-married set that they were soon dubbed "the honeymoon ships." A six-day cruise in the mid-1930s cost $60; a fourteen-day tour, with a stay in a Bermuda hotel, was priced from $114.

Unique presentation and the great variety of delectable dishes offered are special treats for cruisers.

By the late '30s, eleven days from New York to the Caribbean on the *Champlain* cost $140 and fifteen days on the *Kungsholm* started at $182. Longer cruises were still very popular, however, and in a three-month period beginning in December 1938, there were fifteen cruises out of New York that were twenty-four days or longer. Three of these were South American trips, all of which put into Rio for its annual, high-spirited Carnival. One sailing was aboard the most magnificent Atlantic liner of all time, the French Line's *Normandie*. This 83,400-ton beauty had her normal two-thousand-passenger capacity more than halved for her four-week special cruise to southern waters. Her interiors were a press agent's dreams come true: live, caged birds in the winter garden; a dining room done in hammered bronze, glass, and Lalique fixtures; and an eighty-foot-long indoor swimming pool. One passenger, a writer, wrote to friends back in London, "The *Normandie* is just like the very best hotel one can conceive."

Shipboard casinos are one of the myriad diversions for passengers.

The first large liner built especially for the full-time, long, luxury cruises arrived just after the Second World War. She was another Cunarder, the renowned *Caronia* of 1948. She established the high-water mark for cruising in the 1950s. At just over thirty-four-thousand tons or the equivalent of over fifteen miles of freight trains, she generally carried as few as 350 passengers on her long-distance travels (her actual cruise capacity was limited to 600 and absolute maximum was 932, compared to the similarly sized *Nieuw Amsterdam*, which had an over-all capacity of 1,228) and consequently had the overall feel of a floating country club. Onboard standards were impeccable, the tone elegant, and her six hundred staff were hand-picked. The clientele were "regulars," passengers who came year after year. Several ladies "lived" aboard for two and three years at a time, and another had the exceptional record of fourteen years of continuous cruising!

But by the 1960s, when the jets overtook the traditional sea liners, cruising became the mainstay of almost all passenger ship lines and then many new ones as well. Beginning in 1966, Scandinavian ship owners in particular began to see great promise in the U.S. cruise business, particularly out of Miami to the Bahamas and the Caribbean. Specially designed, all-white, all-first-class cruise vessels were launched. Norwegian Caribbean Lines could boast of the *Sunward*, *Starward*, and *Skyward* while later, Royal Caribbean Cruise Lines added a trio of brand-new sister ships: *Song of Norway*, *Nordic Prince*, and *Sun Viking*. And for those longer, more luxurious and expensive trips, the Royal Viking Line was created. By the '70s, cruise ships were traveling to just about every corner of the earth.

Three factors contributed to the growth and popularity of cruising. First, the airlines were no longer a rival, but the friendly transporter of cruise passengers to departure and arrival points abroad. Passengers could join a ship in the Mediterranean, in Australia, even in Antarctica. Second, theme cruises drew many travelers to their first shipboard vacations. There seemed to be titles for just about everyone: sports cruises, gardening cruises, shopping cruises, Hollywood nostalgia cruises, write-a-murder-mystery cruises, even diet and fitness cruises. And third, the original television series *Love Boat*, which ran from

Onboard entertainment runs the gamut from Vegas-style revues to classical music concerts.

A swim, a trip or two down the slide, a lounge in the sunshine, and a hot tub dip with a friend are among the poolside pleasures of a cruise on a Fantasy class ship.

1977 until 1986, and portrayed cruises as fun, uplifting experiences, often with younger, more casual passengers, lured many vacationers to the sea. Each episode of the highly popular show was said to be worth at least $8 million in positive advertising for the entire industry.

It was not until 1988, however, that the first specially built "mega-liner" appeared. The *Sovereign of the Seas*, built for the Miami-Caribbean run, could carry as many as 2,673 passengers. She was the first brand-new cruise ship to pass the seventy-thousand-ton mark. She had vast lido decks, a Las Vegas-style casino, a state-of-the-art health center, and a nightclub perched high above the seas in her funnel. She also had another novelty in shipboard design: the multideck atrium. A new era had begun. More and more cruise liners followed, often grander and more innovative than the ones before. By the early 1990s, the North American cruise industry had reached an extraordinary $6 billion annual level.

The stunning decor of the Rotterdam dining room on Holland America's Statendam *enriches the enjoyment of the world-class cuisine.*

No single company has done more to expand as well as popularize cruising as Carnival Cruise Lines. They began in 1971 with the purchase of one out-of-work, former transatlantic liner, the *Empress of Canada*, which, in a simple makeover of repainting and a few alterations, began plying tropic waters out of Miami. She became the *Mardi Gras*, the first of Carnival's so-called "Fun Ships." Informality, lower prices, and an early emphasis on the sun-seeking singles' market led to success from the very start. A second ship followed in 1976, a third in 1978, and in 1981 the company's first new build, the 36,600-ton, 1,396-passenger *Tropicale*, was launched, less than ten years since the company's inaugural cruise set off. Bigger and bigger-still Carnival ships followed. The eight-ship Fantasy class, each weighing in at just over seventy thousand tons and carrying nearly two thousand, six hundred passengers each, is the largest series of superliners ever built. When the 102,000-ton *Carnival Destiny* appeared in 1996, she was the largest passenger ship yet built.

A setting sun does not signal the end of the day's fun on the Carnival Destiny.

And Carnival has not added just new tonnage, but a corporate empire as well, their "family of companies." One travel writer dubbed them "the Carnival empire." They bought the illustrious Holland America Line with its subsidiary, Windstar Cruises, in 1988 for $625 million. It could not have been a more successful move and soon more and more acquisitions followed: 50 percent of the ultra-luxurious Seabourn Cruise Lines, then Italy's Costa Cruises, shares in Britain's Airtours (which runs ships as well). Most recently, in 1998, Carnival spent $500 million to buy another history-filled shipping company, the 158-year-old Cunard Line. That takeover included what is said to be the most famous ship afloat, the seventy-thousand-ton *Queen Elizabeth 2*, often thought to be the last of the traditional Atlantic ocean liners. Carnival's future plans for Cunard into the new century include a large, very fast liner tentatively known as the "Queen Mary Project." She will continue the traditions of ocean crossings. Carnival has become the biggest passenger ship owner and operator in maritime history.

The lush flora of St. Lucia, Grenadines, welcomes the passengers of Windstar *to this Caribbean port.*

Professionally staffed childcare facilities on the Elation *ensure that parents and children alike enjoy a pleasant cruise experience.*

"Carnival is today's dominant player in the cruise industry because they have excellent business sense," commented George Devol, editor and publisher of the monthly journal *Ocean & Cruise News*, "and this has included diversity in their cruise holdings. They touch upon just about every market—big, mass-market ships as well as select, all-luxury vessels, specialty types and novel ones—and touch all the major cruise areas: from the Caribbean to Northern Europe to Alaska to the South Pacific."

The chairman of a rival cruise line said, "The industry is indebted to

LADIES

H*elpful stewards stand ready to assist passengers on all of today's luxury liners.*

Carnival. They've introduced more vacationers to their first cruise than any other company. Cruising is very addictive. The onboard lifestyle is so appealing. Many later graduate to other ships, perhaps more up-market types, but now these are part of the Carnival family. The range might be from over two thousand passengers on a weekend to Nassau out of Miami on board the *Ecstasy* to a hundred guests on a highly personalized voyage around the Aegean Islands on Seabourn's *Sea Goddess I*. Carnival has it all!"

Carnival's overall success has been in its marketing. It has made cruising in general seem to be fun, more exciting, as compared to the stodgy, only-for-older-and-richer-passengers image of bygone days. They have also appealed to younger, more mainstream, more broad-based vacationers. "Carnival has brought the cruise concept into the mainstream of the public's vacation plans," noted Devol. "Almost single-handedly, they made cruising more affordable and therefore more accessible." By the end of 1998, Carnival owned forty-two passenger ships under six brand names and had no less than ten new ships slated for delivery by 2003.

Altogether, there are over three dozen cruise ships either under construction or on order in shipyards in Italy, France, Germany, and Finland. This is unquestionably the busiest phase of passenger shipbuilding in maritime history. And some of the largest, most lavish vessels are included in this group. There is a trio of 142,000-tonners being built (each with an ice-skating rink on board) as well as an 85,000-ton condominium-style cruise liner. But possibly the most optimistic project of all is the plan for two 250,000-ton, six-thousand-passenger "floating cities." One U.S.-based cruise firm has recently ordered $3 billion worth of new tonnage (ten liners in all) in a single bulk order, while another is spending $2.9 billion for seven ships. The Disney Corporation, which entered the cruise industry with a pair of 85,000-tonners in 1998, is said to be thinking of as many as eight major ships in all. Plans for the famed Cunard Line include a series of all-suite vessels with six and seven hundred berths.

G*uests often become well acquainted with the ship's staff.*

And as cruise design reaches greater heights with the "floating hotel" concept, ships will deliver an even wider range of amenities and services. One new liner has a wedding chapel, a virtual-reality theater, and a southwestern-themed alternative restaurant. One line promises an inline skating track and full basketball courts while another, especially for the Asian routes, will have five restaurants with five different cuisines! The modern, seagoing resort surpasses anything in the annals of ocean liner history.

"Cruising is the best vacation value on this earth," asserted Devol. And more and more vacationers throughout the world are realizing this. The cruise industry is flourishing not only in North America, but in Europe as well. In the United States, five million people took cruises in 1997. One survey indicated that as many as forty-seven million American vacationers say they will "definitely" or "probably" cruise within the next five years.

With more people traveling by ship than ever before, the story of the great ocean liner will continue. This book is dedicated to the "greatest moving objects made by man" and to many of the glorious ports they visit today. Indeed, there is absolutely nothing like a cruise! ⚓

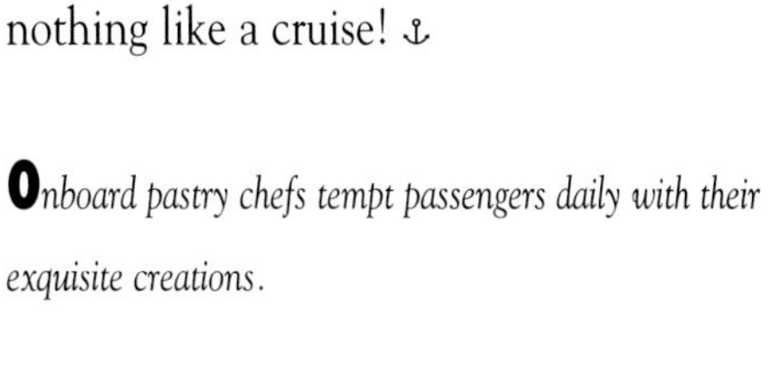

O*nboard pastry chefs tempt passengers daily with their exquisite creations.*

T*he lobby of Carnival's* Elation *is one of several lounge areas that dazzle with their splendor and range of amenities offered.*

Passengers aboard the Seabourn Pride *enjoy the setting sun off St. Thomas in the Caribbean.*

BY BARB AND RON KROLL

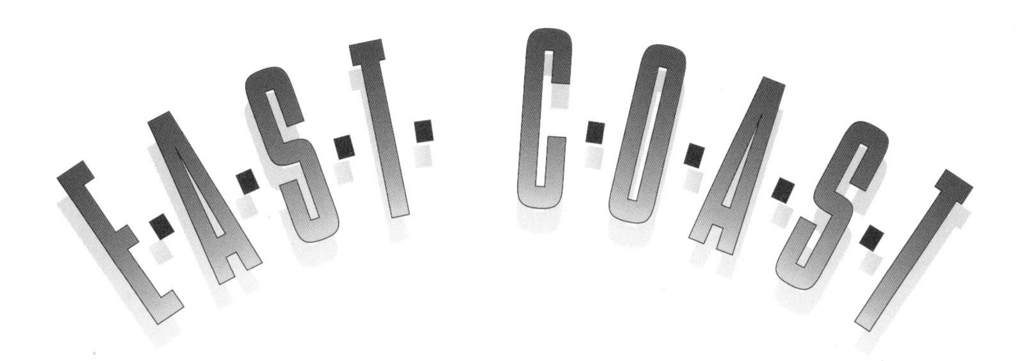

NORTH AMERICA'S ARRAY OF CULTURE

FEW PLACES ON THIS planet offer as much cultural diversity and unbridled fun as the East Coast of North America. A virtual smorgasbord of experiences awaits visitors: basking in the comforts of quaint French Canadian inns, reveling in the hub-bub of cabbies and street vendors below New York's towering skyscrapers, savoring garlicky Cajun delicacies and strummin' banjos in New Orleans, and even coming within jumping distance of jaw-snapping 'gators in Florida. No doubt about it: few regions of the world score higher marks for adventure, romance, culture, and shopping.

"Let the good times roll!" is the unofficial motto of New Orleans, where Mardi Gras brings nonstop revelry, decorated floats, and masqueraders dancing through the streets during the week before Ash Wednesday. Music pervades the city 365 days a year. Jazz medleys flow from the open windows of bars and restaurants throughout the French Quarter while pedestrians rove the streets as if they were at some gigantic outdoor cocktail party. Many visitors end up at Preservation Hall, where Louis Armstrong emulators pour their hearts and souls into jazz classics. Even in the wee hours of the morning, the muted wailing of trumpets marks the dying hours of all-night jam sessions.

New Orleans's genteel elegance often puts visitors in the mood for love. Flowers hang from wrought iron balconies and flickering gas lamps line the narrow streets. Pastel facades hide inner courtyards filled with sweet-scented flowers and cascading fountains—perfect places for romantic rendezvous on warm sultry nights. And what better place to spend a New Orleans morning than at the French Market, sipping café au lait and nibbling on beignets (freshly baked doughnuts liberally dusted with powdered sugar), watching paddle wheelers ply the Mississippi River—a reminder of the days when they carried cotton rather than tourists. The pleasant clamor of their steam calliopes is as much a part of New Orleans as saxophone melodies created by street-corner musicians.

BERMUDA
BOSTON
CHARLESTON
KEY WEST
MIAMI
MONTREAL
NEW ORLEANS
NEW YORK
PALM BEACH
QUEBEC CITY
TAMPA

The seven spikes on the crown of the Statue of Liberty represent the seven continents and the seven seas—all well-traveled by stately ships that carry cruisers throughout the world.

Steeped in history and culture, New Orleans is full of surprises and fascinating sights. St. Charles Avenue's streetcar, the oldest continuously running street railway in existence, offers stunning views of the Garden District's stately mansions. For a much different sort of scene, visitors flock to the homes of the dead—the city's famous "high-rise" cemeteries. Since much of New Orleans lies below sea level, the aboveground graves thwart flooding. Perhaps the most famous tomb belongs to Marie Laveau, the voodoo queen. Her former home is now the Voodoo Museum filled with creepy occult artifacts, good luck charms, love potions, and a wide array of voodoo dolls for hexing one's enemies.

The city's other magic charm is food. Seafood is a staple of New Orleans cooking, which means you can feast on basketfuls of boiled crawfish and platters of deep-fried oysters, shrimp, catfish, and soft-shelled crab. Or you can savor tantalizing Cajun and Creole specialties like gumbo and jambalaya. And to satisfy that sweet tooth, you might want to sink your teeth into buttery pecan pralines.

Florida lies only about 120 miles east of Louisiana, but the ambience of these two Dixie states could not be more divergent. In many respects Florida is like an eclectic art gallery, offering completely different pictures in every direction. Look to your left and you'll see a Florida that is thoroughly gaudy, an endless expanse of cheap motels, neon signs, and cheesy theme parks. Look to your right and witness the sublime, especially down around the Keys where the warm breezes of the Caribbean consort with a laid-back lifestyle. Or look behind you to experience the downright savage: the rolling thunder of Daytona or the primal grunts of gators in the Everglades. Florida is all these things and more, one of North America's most fascinating destinations.

Florida's cruise ports are highly unique. Key West has always been eccentric, laid-back, and gay, a place where just about anything goes—from nautical nuptials with newlyweds and guests dressed in scuba gear, to wacky headstones in the municipal cemetery. One tomb bears the words, "I told you I was sick." Another, engraved by a widow, reads, "At least I know where he's sleeping tonight."

Adventures await all along the Florida Keys: deep-sea fishing, swimming with dolphins, and snorkeling along underwater trails. You might even stumble onto sunken treasure. Many moons ago, the archipelago was a haven for pirates waiting to pounce on Spanish treasure ships on their way home from Mexico and Cuba. Hurricanes and sunken reefs also took their toll on Gulf Stream shipping, with hundreds of ships going to watery graves. Some of the wrecks have yielded sunken riches—gold doubloons, silver ingots, and precious jewels.

Even the horses dress to impress in New Orleans.

Once you've sharpened your appetite on the salty air, dig into the treasure trove of seafood that awaits at Key West restaurants. Chowder, conch fritters, and Key lime pie are among the local gems. At Ernest Hemingway's house, descendants of his six-toed cats still prowl the premises. You can even stop for a drink at Sloppy Joe's, one of Hemingway's favorite bars.

While Key West induces blissful languor, Miami offers a more upbeat tempo. Cars parade along palm-lined Ocean Drive. Rollerbladers zip along pink sidewalks. Alfresco cafés overflow with diners. Bronzed lifeguards survey sunbathers from the shade of their art deco beach huts. The wide white strands are alive with glistening joggers and models posing for fashion shoots. After sunset, neon illuminates America's finest collection of art deco buildings, many of them now listed as historic landmarks. Couples stroll hand in hand, browsing art galleries and boutiques along the leafy boulevards.

Unique daytime activities also abound around Miami. Spanish-speaking Little Havana holds many places to discover: Cuban restaurants, cigar shops, sidewalk domino competitions, and botanicas that sell amulets and spiritual consultations. You can also immerse yourself in

The wrought iron of the French Quarter is but one element that makes New Orleans such a romantic port to visit.

THE SOUNDS OF

NEW ORLEANS

New Orleans and the surrounding bayou country have given birth to several distinctive musical forms, including Dixieland jazz and Cajun zydeco. The city was the starting point of a number of illustrious music careers, bygone legends like Louis Armstrong, Jelly Roll Morton, and King Joe Oliver, as well as modern stars like Buckwheat Zydeco, Queen Ida, Clifton Chenier, Wynton Marsalis, and Harry Connick Jr. Preservation Hall on St. Peter's Street is the best place to catch traditional jazz. Other popular music spots include the House of Blues, Benny's Bar, and Tipitina's, which showcases local zydeco groups. ⚓

P*ort appeal rating*

Adventure	★★★★½
Entertainment	★★★★½
Romance	★★★★½
Cuisine	★★★★★
Shopping	★★★½

the primordial ambience of the Everglades, just a short drive from Miami. As you canoe or houseboat through mangrove-lined channels, herons soar overhead and alligators use their periscopic eyes to hunt for prey. Gentle Florida manatees sometimes offer snorkelers delightful surprise visits.

If Miami is largely shorts and sunglasses, Palm Beach is all about diamonds and twenty-four-carat gold. For nearly a hundred years this stately coastal enclave has been the haunt of the rich and famous, a place where "Eastern money" came to winter or retire. Whether you consider Palm Beach an East Coast version of Beverly Hills or a tropical take on Newport, the overall effect remains the same: an ultra-exclusive colony populated by citizens who never need to ask the price of anything.

Chauffeur-driven limousines and flashy European sports cars cruise the avenues. Pink and purple bougainvillea cascade over sparkling white walls that surround mansions large enough to

pamper Godzilla. Cobbled passageways open into courtyards brimming with potted plants, marble statues, and tiled fountains. There's even a drinking fountain for the city's pampered pooches—but even they wouldn't dare approach one of the polished chrome fire hydrants. If you can pry yourself away from the fabulous beaches and golf courses, consider some of the other forms of local entertainment: greyhound racing, a polo match, or a croquet tournament. To paraphrase F. Scott Fitzgerald, the rich really are different, especially in Palm Beach.

The sultry beauty of the Florida Keys has inspired generations of visitors, including such famed American writers as Tennessee Williams.

THE BUSTLE OF

KEY WEST

Somewhere along the line, nondescript Key West became funky Key West, and now the city at the end of the Florida Keys is a huge tourist destination. Day or night, busy Duval Street is the focus of the city's entertainment and shopping. Bars (including the one Ernest Hemingway frequented) do a booming business, and the shops carry every conceivable T-shirt message. The city's most famous event is the Sunset Celebration at the Mallory Dock, a spontaneous carnival that brings out assorted buskers, flame-eaters, shills, and street food vendors as well as hundreds of curious visitors. ⚓

Port appeal rating

Adventure	★★★★
Entertainment	★★★½
Romance	★★★★
Cuisine	★★★½
Shopping	★★★

THE TRENDS OF

TAMPA

With miles of white Gulf beaches, a major theme park (Busch Gardens), three professional sports teams, and a whole lot of history, the Tampa area region finds it easy to entertain visitors. Combining history with nightlife is Ybor City, the hottest evening hangout in the region, where trendy cafes and bars occupy warehouses once used by Cuban and Italian immigrants to make cigars. Disney World and other Orlando theme parks lie only a couple of hours away by car. ⚓

Port appeal rating

Adventure	★★★½
Entertainment	★★★½
Romance	★★★
Cuisine	★★★
Shopping	★★★

Handpainted ceramic tiles and moorish influences are part of the facination and historic charm of Florida.

Right: *Worth Avenue, with its shaded arches, offers one of the premier shopping experiences in Palm Beach.*

Overleaf: *The regal* Carnival Destiny *sails out of Miami, carrying cruisers to exotic locales.*

THE DAZZLE OF

PALM BEACH

The regal resort city of Palm Beach is something to see. Huge mansions from the Roaring Twenties hide behind well-manicured hedges. Tony hotels offer elegant quarters and impeccable service. Chauffeured Rolls Royces line Worth Avenue, where shops, jewelers, and couturiers, offer the finest of shopping with full amenities including a sidewalk fountain for Madame's thirsty dogs. ⚓

Port appeal rating

Adventure	★★★½
Entertainment	★★★½
Romance	★★★½
Cuisine	★★★½
Shopping	★★★½

THE TIDES OF

MIAMI

The world's biggest cruise port started out as a winter resort, then grew to become an international commerce center. Today, cruise ships dock in the shadow of downtown Miami's skyscrapers in what has to be one of the most scenic harbors on earth. Across Biscayne Bay lies fabled Miami Beach, whose sands have seduced vacationers from all over the globe. There, too, is art deco South Beach and Ocean Drive, the hottest resort scene in the country, where visitors sip drinks at sidewalk cafes facing the ocean as models and wannabe movie stars roll by on inline skates. Just twenty-five miles north in Fort Lauderdale is the major cruise port of Port Everglades, whose harbor fills with liners every winter. ⚓

Port appeal rating

Adventure	★★★★
Entertainment	★★★★
Romance	★★★½
Cuisine	★★★★
Shopping	★★★½

Carnival Destiny

THE CHARM OF

CHARLESTON

South Carolina's barrier islands provide both shelter from Atlantic storms and wonderful escapes from urban life. Hilton Head is a sportsman's mecca where golf and tennis reign supreme. Kiawah, a resort community, is a nesting place for loggerhead turtles where beachfront homes sport subdued outdoor lighting in deference to the sensitive turtles. The Cape Romain Group is a rich wildlife refuge where 250 different species of birds frolic in marshes, lagoons, and woodland. Most of the islands supported cotton, indigo, and rice plantations until the Civil War, and some of the more rural areas still reflect life in the antebellum South. ⚓

Port appeal rating

Adventure	★★★★
Entertainment	★★★½
Romance	★★★★
Cuisine	★★★★
Shopping	★★★½

Tampa lies on the other side of the Florida Peninsula (the balmy Gulf Coast) and the opposite end of the cultural spectrum. The Bay Area—which encompasses Tampa, St. Petersburg, Clearwater, Palmetto, and a dozen other cities—is solidly middle class and middle income. Residents are more likely to follow baseball or football than polo, and are probably more comfortable with a bottle of cold beer than a glass of expensive champagne. But that doesn't mean Tampa is without its charms. Busch Gardens doubles as a beer garden and African safari park with over thirty-four hundred animals. For lovers of marine life, the Florida Aquarium depicts the state's fresh- and saltwater habitats. If you want to experience the fury of a tropical storm without actually living through the terror, visit the Gulf Coast Hurricane Exhibit at the Museum of Science and Industry where you can lean into howling seventy-five-mile-an-hour winds, hair flying and clothes flapping, while you dodge flying objects like chewing-gum wrappers, eyeglasses, and earrings.

Key West sits like a pearl at the tip of the Florida Keys.

Ybor City, another Tampa Bay area community, offers a whole different sort of blast. In 1900, twelve thousand cigar makers filled two hundred factories here, rolling panatelas and coronas for the stogie smokers of America. Ybor is a place where red brick streets front Spanish-style buildings, their white stucco, intricate grillwork, and blue and gold floral wall tiles sparkling in the sunlight. Laughter and cigar smoke emanate from clubs where members play dominoes and cards. The pungent aroma of tobacco overwhelms your nostrils as you enter a former cigar factory where workers to this day still roll and cut stacks of the cured brown leaves. More enticing scents of roast pork and strong coffee draw you into small cafés for inexpensive but tasty Cuban food.

Like an old Johnny Mercer tune, Charleston comes at you really slow and sweet. The queen of the Carolina coast offers a trip back in time to an era when manners were more refined and people took time to smell the roses . . . or rather, the magnolias in this case. There are few more historic cities in North America. Founded in 1670 by British colonists, most of them indentured servants, Charleston soon grew into a thriving port that shipped cotton, indigo, flax, and other southern products. On a more infamous note, the port also handled slaves, operating as one of the largest flesh markets in the Americas. Charleston flourished through colonial times, the antebellum period, and even the nefarious Reconstruction after the Civil War.

After leveling Atlanta, Sherman was hell-bent on torching Charleston. But the city was spared destruction because it surrendered to the Union navy rather than the vengeful general. That decision continues to echo through the ages because Charleston is one of the few southern cities that preserves its wonderful antebellum charm. "Charleston is a never ending feast for the senses," wrote one local author, "with history as the main course." Indeed, the past is what makes the city so special today—a jubilee of mansions and plantation houses, formal gardens and fortifications, and the country's oldest museum.

Nearly six hundred miles offshore from the Carolinas lies the island of Bermuda, pretty and properly British. Ice-cream-colored homes with sugar-cube-white roofs, bountiful flowers tumbling over stone walls, pink sand beaches, and aquamarine seas accent the twenty-one-mile-long island. At the Supreme Court, judges and attorneys still wear gowns and wigs. Ships usually dock alongside in the capital city of Hamilton. Just steps away, shops sell imported merchandise like British tweeds, Irish linen, and Scottish cashmere. There are no rental cars on the island, which helps preserve the pristine environment, but it's fun to put-put around the island on scooters. If you plan to sample Dark 'n' Stormies, the traditional drink of Black Seal rum mixed with ginger beer, Bermuda's pretty pink buses offer a safer alternative to driving.

Barracuda, sharks, and moray eels live in the aquarium. For a closer glimpse of the island's marine life, even nonswimmers can stroll underwater, wearing old-fashioned lead helmets attached to air hoses. Colorful fish flick their fins tauntingly in front of your nose then vanish into coral gardens, leaving only air bubbles behind. Back on dry land, the fragrance of jasmine, oleander, and frangipani scent the air. The Bermuda Perfumery

Charleston's rich heritage enhances a stroll along her walkways.

A cruise ship celebrates arrival in Miami.

Autumn runs riot in Salem, Massachusetts.

makes and sells fragrances that will trigger memories of Bermuda long after your visit.

Ships also visit St. George's, the first settlement on the island and the second English town established in the New World. Multiscreen slide presentations in the town hall bring 375 years of Bermuda history to life. So does the town crier who rings a bell and bellows, "Oyez! Oyez!" to announce reenactments of public punishment in King's Square. Only much-loved cricket games draw bigger crowds.

Be they immigrants bound for Ellis Island or luxury cruise ship passengers on their way to Broadway, no one ever forgets the first time they sail into New York Harbor; the Statue of Liberty off to port and the skyscraper-studded skyline of Lower Manhattan off to starboard leave an indelible impression on all who steam up the Hudson.

It's impossible to see everything the Big Apple has to offer in one trip, or a dozen trips, for that matter. First-timers usually do the essentials like the Empire State Building and the Metropolitan Museum of Art. On subsequent visits you can delve into other attractions like South Street Seaport with its historic sailing ships. Special interest tours allow you to savor soul food and jazz, discover celebrity homes, or glimpse behind-the-scenes goings-on at NBC Studios and the frenzied nerve center of the New York Stock Exchange.

Wherever you go, excitement is the hallmark of the city, be it Broadway with its dazzling shows, Times Square ablaze with lights, or city streets filled with cacophonous traffic and jostling crowds. New York is a city of superlatives—encyclopedic collections in its museums and galleries, cloud-touching buildings, and outrageously expensive shops. You can eat your way around the world in New York City, from Peking duck in Chinatown to veal parmesan in Little Italy. There are no shortages of romantic getaways, either. You can slow down the pace by clip-clopping through Central Park in a horse-drawn carriage. Or dine and dance on top of the World Trade Center with the world at your feet, glimmering with lights like diamonds on black velvet. Whatever you want, you can find it in New York City. Anything!

Boston is another city of many personalities. The aroma of hot dogs and the cheering of Red Sox fans at Fenway Park will convince anyone that it's a city of sports lovers. In Cambridge, home to Harvard, America's first university, the mood is more cerebral. Bookstores abound. So do shops selling preppie tweed jackets with leather elbow patches. But Boston also exudes history. Strolling along the Freedom Trail, you'll encounter sixteen sites that are each significant to American history, from Paul Revere's house to the Old South Meeting House where citizens gathered in 1773 to protest a tea tax and launch the Boston Tea Party.

Charming gaslights, cobbled streets, and nineteenth-century brick townhouses characterize Beacon Hill, where antique shops beckon with "finds." Shopping at Faneuil Hall is completely different. Jugglers, mimes, and musicians entertain as shoppers search for scrimshaw and pewter displayed on colorful pushcarts. The scent of fresh flowers mingles with aromas of Greek shish kebabs, Chinese stir-fries, and Canadian bacon. Italian treats like cannoli and cappuccino are found in the cafés along the narrow twisted streets of the North End. Strings of fennel and garlic-scented sausages hang in the windows of neighboring shops above pyramids of red plum tomatoes. Ravioli presses and espresso machines make great souvenirs.

The high-rise skyline of Lower Manhattan is recognized the world over.

Nobody on fishhook shaped Bermuda lives more than a mile from the sea.

THE BEAUTY OF

BERMUDA

What is the oldest legislative body in the western hemisphere? The colonial assembly in Bermuda, established in 1620—the same year that English settlers were just getting started at Jamestown, Virginia. Nearly four centuries later, the tiny archipelago (320 islands) is still part of the British Empire, with a governor general who answers directly to the queen. Bermuda is often lumped together with the English-speaking Caribbean isles, but it actually lies much farther north, 580 miles off the coast of North Carolina, which means milder weather and much less danger from hurricanes. ⚓

Port appeal rating

Adventure	★★★★
Entertainment	★★★
Romance	★★★★½
Cuisine	★★★½
Shopping	★★★½

Other cities may now have taller structures, but nobody can match the sheer artistry of New York's famous skyscrapers. The Flatiron Building was the first of the city's giants, erected in 1902 and decorated with Florentine Renaissance motifs. Eleven years later, the Woolworth Building shot into the sky above Lower Manhattan, a modern Gothic marvel with gargoyles, frescoes, and stone lacework. That was the world's highest structure until 1930, when an art deco masterpiece called the Chrysler Building soared over Midtown. A year later, the Empire State Building grabbed the tallest title (and then nearly lost it to King Kong). ⚓

Port appeal rating

Adventure	★★★★☆
Entertainment	★★★★★
Romance	★★★★
Cuisine	★★★★★
Shopping	★★★★☆

Cunard's QE2 sails below the twin towers of the World Trade Center in New York City.

CUNARD

THE HERITAGE OF

BOSTON

Harvard makes a good day-trip from Boston. America's most renowned university is easily explored on foot, starting from Harvard Yard with its historic buildings. Ivy-clad Massachusetts Hall (erected in 1720) is the oldest structure on campus, while the granite University Hall (1816) contains the administration. Widener Library, named after a Harvard grad who went down with the *Titanic*, is the world's largest college library. Harvard also offers nine museums, including three art galleries and four natural history collections. The Fogg Museum alone boasts twenty-seven works by Rodin, two hundred by Rembrandt, and three hundred by Durer. ⚓

P*ort appeal rating*

Adventure	★★★★
Entertainment	★★★★
Romance	★★★½
Cuisine	★★★★
Shopping	★★★½

Ethnic Boston is rivaled only by maritime Boston along the waterfront where its seafood restaurants offer clam chowder, catch-of-the-day, and one of Boston's local brews. After all, the city is home to the original Cheers bar. Boston also offers its share of romantic pursuits. You can cuddle into a swan-shaped paddleboat for a cruise around the Public Garden's tranquil pond or rent bicycles to pedal through five scenic parks on a trail called the Emerald Necklace. Romance, history, culture, sports—Boston offers them all in abundance.

Canada's eastern ports of call are truly affable places. In Halifax, the mayor invites visitors for afternoon tea on summer weekdays. While dining on succulent lobster, scallops, and mussels or visiting pubs, you'll meet friendly locals who may invite you to a *ceilidh* (pronounced *kay-lee*), where fiddlers play Celtic tunes until early morning. Perched on a windswept hill, Halifax's star-shaped fortress keeps a wary eye on the harbor. The city is made for walking. You can stroll the floral-scented Public Gardens, or wind your way down to the sea at Point Pleasant Park. The salty tang of seaweed, squawking seabirds, and rhythmic lapping of the waves leave no doubt that this is a maritime city.

The Rocky coast of Maine is dotted with fishing villages

In Quebec City, the joie de vivre occurs around restaurant tables with plenty of free-flowing wine and mouthwatering tourtiere (meat pie) and tarte au sucre (sugar pie). Quebec has its own massive citadel, built by the British in 1832; you can stroll along its ramparts for nearly three miles. The only walled city north of Mexico, Quebec was also the first North American city on the United Nation's prestigious World Heritage list—along with such renowned attractions as Egypt's pyramids, the ruins of Angkor Wat, and the Statue of Liberty.

Quebec's Upper Town sits on a rocky cliff 250 dizzying feet above the St. Lawrence River. Eleven staircases link it to the Lower Town, which nestles between the cliff and the river. Breakneck Stairs lead down to Place-Royale, the oldest town square in all of North America. From there, an inexpensive ferry ride takes visitors to Levis on the south bank. Views of the city, dominated by the massive, turreted Chateau Frontenac, are breathtakingly romantic, especially in the morning when sunlight gilds the city. A funicular transports pedestrians to the Frontenac, which is to Quebec City what the Eiffel Tower is to Paris.

Both French and English cultures permeate Canada's provinces.

Even farther up the St. Lawrence is astonishing Montreal, the world's second largest French-speaking city after Paris. As Expo '67 and the 1976 summer Olympics confirmed, this is very much a city that looks toward the future without forgetting its rich past. Sure, Montreal boasts its fair share of Gothi-inspired churches and ivy-shrouded universities. But flashy steel-and-glass skyscrapers dominate the downtown area and modern art flourishes in public places. While the nightlife scene is sophisticated, a more vital sign of the city's heartbeat is the annual jazz festival, which spills over into the streets and parks.

From New Orleans to Montreal, the ports of the North American East Coast offer cruise passengers a unique combination of ambience, history, cuisine, and romance. Each visit invariably whets the appetite for future discoveries. ⚓

Halifax sounds a hearty welcome.

Boston, birthplace of the American Revolution, is the home of the USS Constitution.

THE ENCHANTMENT OF

QUEBEC CITY

Romance pervades the Old City. The streets are narrow and lined with pretty stone buildings, the scent of freshly baked bread wafts from the patisseries, and villagers gather at sidewalk cafés to discuss the state of the world over cafe au lait. Equally romantic is the landmark Chateau Frontenac Hotel on a bluff overlooking the St. Lawrence River, from which couples can go touring in a horse-drawn carriage. ⚓

P*ort appeal rating*

Adventure	★★★★
Entertainment	★★★½
Romance	★★★★½
Cuisine	★★★★½
Shopping	★★★½

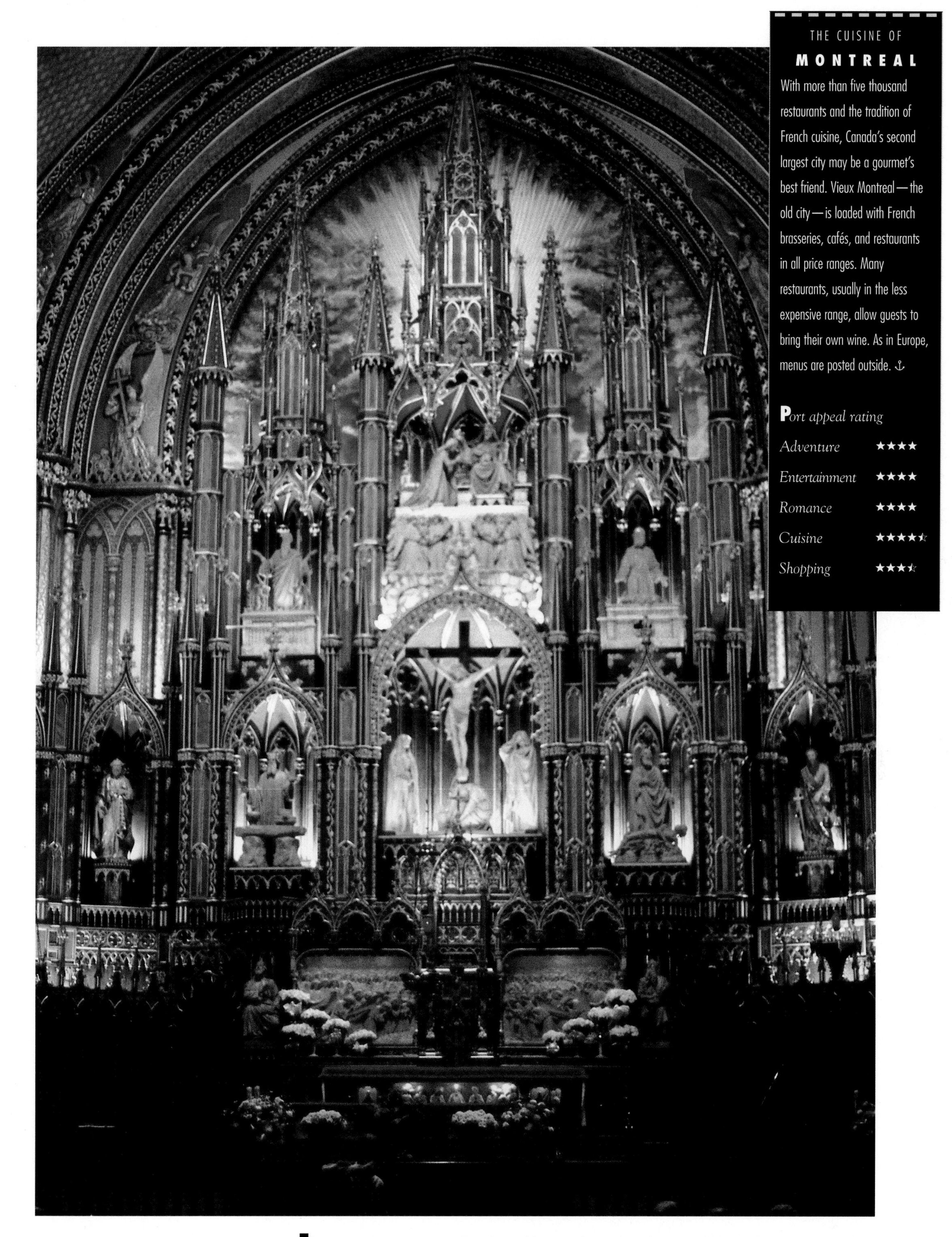

The imposing Notre Dame Basilica, designed by an Irish American, has stunned Montreal's visitors since 1829.

THE CUISINE OF

MONTREAL

With more than five thousand restaurants and the tradition of French cuisine, Canada's second largest city may be a gourmet's best friend. Vieux Montreal — the old city — is loaded with French brasseries, cafés, and restaurants in all price ranges. Many restaurants, usually in the less expensive range, allow guests to bring their own wine. As in Europe, menus are posted outside.

Port appeal rating

Adventure	★★★★
Entertainment	★★★★
Romance	★★★★
Cuisine	★★★★½
Shopping	★★★½

The hotel Chateau Frontenac has dominated the Quebec City skyline for over a century.

BY JOSEPH R. YOGERST

C·A·R·I·B·B·E·A·N

THE TREASURED ISLANDS

"A LOT OF PEOPLE catch the treasure bug," said a man who has roamed the Caribbean for nearly twenty years, poking through shipwrecks and pirate caves. "And just about every island I know has its stories about lost treasure."

Indeed, there is a possibility, no matter how remote, that you just might chance upon undiscovered riches in the Caribbean. Picture yourself on some secluded beach. Suddenly you see a glimmer in the water—a coral-encrusted coin baring the likeness of a long-forgotten sovereign, washed ashore by a recent storm. The copper may be no more than a hundred years old, but that doesn't matter. You've found your first Caribbean treasure and you want more.

One of the most popular spots for underwater adventure is the Virgin Islands—both the American and British halves—which offer a dozen fascinating spots within a short cruise of one another. In the secluded bays of St. John you can snorkel among eagle rays, baby barracuda, delicate fan coral, and queen angelfish. On the outer edge of Salt Island you can poke through the submerged remains of the "Caribbean's Titanic"—a British steamer called the RMS *Rhone* that went down in a hurricane in 1867 with the loss of more than three hundred lives. And over on cactus-studded Norman Island, you can explore underwater caves that inspired the definitive book on pirates and plunder—*Treasure Island* by Robert Louis Stevenson.

Scholars believe that Stevenson based his fictional tale on a real-life pirate adventure that took place in 1750. When a Spanish galleon wrecked on the Carolina coast, two American sea captains—under the ruse of aiding the Spaniards—absconded with most of the gold. Seeking shelter in the Virgin Islands, the Americans were swindled by a certain Captain Norman, who buried the loot on a desert island that now bears his name. Norman and the Carolina captains were later caught by vengeful Spaniards, but it appears they walked the plank without spilling the beans on the treasure's exact

ARUBA
BARBADOS
CURAÇAO
GUADELOUPE
HAVANA
MARTINIQUE
OCHO RIOS
SAN JUAN
ST. CROIX
ST. LUCIA
ST. MARTIN
ST. THOMAS

The colors and costumes of Carnival captivate revelers in Charleston, the tiny capital of Nevis.

location. Rumors to the contrary, there's no evidence that anyone has ever found Captain Norman's cache. Which makes a trip to Norman Island all the more intriguing—you may just be the one who finally discovers the long-lost treasure.

Treasure Island is perhaps the most famous tale about the West Indies, but it's by no means the only one. The region's literary trove is nearly as rich and diverse as the human history. Native writing runs a broad spectrum from the hard-edged coming-of-age stories of Antigua's Jamaica Kincaid (*Annie John*) to the epic poetry of Derek Walcott (*Omeros* and *The Bounty*)—the St. Lucian who won the Nobel Prize for Literature—to the magical realism of Cuban novelist Alejo Carpentier (*The Kingdom of This World*). The islands have also proved fertile ground for myriad American and European authors. Graham Greene took his keen political and social commentary to new heights in Caribbean-based works like *Our Man in Havana* and *The Comedians.* For pure satire there's nothing as funny as Herman Wouk's *Don't Stop the Carnival*—which singer Jimmy Buffet recently transformed into a Broadway musical comedy. Ian Fleming, a longtime Jamaica resident, set one of the very first James Bond adventures in the West Indies. And anyone who doesn't think the Caribbean can be a terrifying place should consider the underwater horror of Peter Benchley (*The Deep* and *The Island*) and the hair-raising colonial tales of H. G. de Lisser (*White Witch of Rosehall*).

Stingray City off the north coast of Grand Cayman affords visitors close-up views of marine life.

There's a good argument to be made that the modern bards of the Caribbean are the singer-songwriters who have spread the music of the region around the globe. From the gentle wail of steel drums to the earthy cadence of folk songs to that brazen adrenaline rush called salsa, the West Indies have given birth to half a dozen major musical styles.

This eruption of talent was sparked by the melding and merging of various European cultures with African slave tunes and the much later infusion of American jazz, blues, and rock. Caribbean music works on a number of different levels: a medium of love and hate, celebration and rebellion, the fundamental way that islanders express their feelings to the world at large. From Bob Marley to Jimmy Cliff to Peter Tosh, Jamaica is a hotbed of reggae and all its various accoutrements—dreadlocks, ganja, and knitted tams. Many people time their visit to Montego Bay to coincide with Reggae Sunsplash, the region's oldest and most extravagant music festival. Trinidad explodes with calypso music during carnival in February, a melody of steel drums and flamboyant costumes that float through the crowded streets of Port of Spain and other towns. Merengue and salsa are staples of Latin ports like San Juan and Santo Domingo, while French-speaking islanders still argue over whether the beguine began in Guadeloupe or Martinique.

Caribbean cuisine is also eclectic, the product of the same cultural fusion that created West Indian music. In fact, eating is one of the most underrated pleasures of any trip to the region. Creole cooking reaches its savory summit on the French isles, especially Guadeloupe, where local seafood is combined with spicy African seasoning and subtle French sauces. The edibles on Aruba and some of the other Dutch possessions carry a definite hint of Holland, dishes like keshi yena (Edam cheese stuffed with meat, onions, pickles, raisins, and olives) and rijsttafel (a rice-based feast imported from the Dutch East Indies). Puerto Rico and other Spanish-speaking islands offer mouthwatering Iberian dishes like bacalao (codfish stew), empanadas (turnovers filled with spicy meat, seafood, or vegetables), and roast suckling pig. Meanwhile, the British raj endures in Barbados where afternoon tea with scones, clotted cream, and cucumber sandwiches is still de rigueur.

From the lovely Anse Chastanet beach in St. Lucia, beachcombers can view volcanic peaks, Petit Piton and Gros Piton.

THE CHARMS OF

ST. LUCIA

Travelers often say there is a certain poetry about the Caribbean, but nowhere is this more evident than St. Lucia, second largest of the Windward Isles. Derek Walcott, the only West Indies writer to win a Nobel Prize for literature, was born in Castries, the island's charming little capital; few would doubt that St. Lucia's eclectic culture influenced his words. The island changed hands between France and Britain no less than fourteen times. And although it was British for two centuries leading up to independence, there's still plenty of Gallic influence in the local cuisine, language, and architecture. ⚓

Port appeal rating

Adventure	★★★★
Entertainment	★★½
Romance	★★★★
Cuisine	★★★
Shopping	★★½

THE WONDERS OF

MARTINIQUE

It's not quite like strolling down the Champs-Elysees, but in many respects the bustling port of Fort-de-France is a miniature version of Paris, albeit with a tropical touch. Martinique has been French for nearly four centuries and is now a department of the mother country, a status similar to Hawaii's statehood. The island has always had a strong affiliation with French womanhood, from royal splendors like Empress Josephine (who was born here in 1763) to the topless beauties who frequent the white-sand beach at Pointe du Bout. Martinique's shopping also reflects the best of French femininity: Hermes scarves, Baccarat crystal, and Chanel perfume. ⚓

Port appeal rating

Adventure	★★★½
Entertainment	★★★
Romance	★★★★
Cuisine	★★★½
Shopping	★★★

Seafood is abundant around the region. Barbados is famous for its flying fish, served steak style or wedged between two pieces of bread. Conch of one kind or another—fritters, steaks, salad—is one of the local specialties in Cayman. The stuffed crab on St. Martin is divine, and you can't pass through the Virgin Islands without a taste of herring gundy. Jamaica offers a veritable smorgasbord of tropical delights: saltfish and akee, super-spicy jerk pork cooked over a barbecue pit, rice and kidney beans cooked in coconut oil, fried breadfruit or plantain drenched in melted butter, and a marinated fish dish called escovitch. And rum. All sorts of rum and rum liqueurs produced by distilleries around the island.

Divers along Grand Cayman's North Wall experience one of the world's most fascinating dive spots.

Reggae and rum aside, Jamaica is a good example of the Caribbean's diversity, how a single island can be many things to many people. There is a well-trodden tourist route across the island's north shore. Stumble and tumble your way up beautiful Dunn's River Falls. Sail the deep blue waters off Montego Bay. Sink into shameless hedonism in Negril. But there is also a "secret" Jamaica, an island that most people, even many Jamaicans, never encounter. Search for the elusive Caribbean crocodile in the murky waters of the Black River. Ride horses across the lush tropical hills and dales of an old colonial plantation south of Ocho Rios. Feast on homemade mango and coconut ice cream in the shade of a huge banyan tree in the courtyard of Devon House in Kingston. Hike through aromatic coffee plantations to the crest of Blue Mountain Peak to watch sunrise over the Caribbean.

Peel back another layer of Jamaica's eclectic heritage and you find even stranger things, perhaps a few skeletons in the closet. Like Seaford Town, nestled in the lush green mountains between Kingston and Montego Bay. It's still populated by the descendants of slaves. Not black slaves, who once worked sugar cane fields and molasses factories all around the Caribbean, but white indentured servants recruited in Germany in the 1830s. Today there are half a dozen German communities perched in the rugged highlands of central Jamaica, one of those little quirks of fate—call it "historical treasure"—that makes it such a bewitching island.

Antiguan hospitality extends all the way down to the shore.

Barbados is another island that can change dramatically from one mile to the next. The west coast between Prospect and Speightstown is about as glitzy as it gets: champagne and caviar at the polo club, million-dollar mansions along the waterfront, and the supersonic Concorde winging in twice a week from London. The highland Scotland District is the spitting image of its Gaelic namesake: rolling green hills grazed by sheep and russet-colored cattle. The area is bleak and windswept like the Scottish moors, a dramatic landscape that's ideal for hiking and horseback riding. And then there is Bridgetown, the island's bustling capital, that is the epitome of a colonial town in the Caribbean. A statue of Lord Nelson looks down upon the local Trafalgar Square, which predates the London version by thirty-six years. Bridgetown cops strut their stuff in chipper British uniforms and judges still preside in flowing black gowns and white wigs. The local parliament—firmly entrenched in the Westminster system—is the third oldest elected assembly in the western hemisphere.

But there are other parts of Barbados that harken back to ancient ancestor times, that seem to have changed little over the last ten thousand years. One such place is Welchman Hall Gully, the remnant of a massive limestone cavern that collapsed in prehistoric times. The rocky ravine is now filled with all sorts of tropical plants—macaw palms with their spiky trunks, breadfruit and cannonball trees, red gingers, white begonias, a dozen different types of orchid, and great fig trees hung with moss that give the island its name (*barbados* means *bearded men* in Portuguese). Welchman offers the Caribbean as it must have looked and felt before the advent of human settlement.

Victorian clock towers are part of the architectural mélange and charm that await visitors to Trinidad's Port of Spain.

Legends of treasure also persist on the Turks and Caicos Islands (TCI),

Martinique offers fine shopping and leisurely strolls in the cooling shade beneath coconut palms.

Singers welcome visitors with Jamaican music on the steps of Rosehall near Montego Bay—a favorite stop for cruisers, and home to a notorious ghost.

THE OFFERINGS OF

OCHO RIOS

Ocho Rios rocks with reggae, rum, and radical waters. One of the world's most beautiful sights—and Jamaica's most renowned landmark—Dunn's River Falls, majestically cascade down to the sea just outside Ocho Rios. Other than rafting down the Martha Brae River or diving along the offshore reef, you can wander along nearby Montego Bay's Doctor's Cave Beach. Jamaica's beautiful North Coast has been home to such famous folks as Ian Fleming, author of the James Bond books, playwright Noel Coward, and actor Errol Flynn. ⚓

Port appeal rating

Adventure	★★★½
Entertainment	★★★
Romance	★★★½
Cuisine	★★★
Shopping	★★½

an unsung British colony at the eastern end of the Bahamas. Ponce de Leon passed through TCI on his quest for the fountain of youth and there is strong evidence that Christopher Columbus had his first landfall here rather than San Salvador. In days gone by, pirates knew the archipelago for its excellent hideouts. Cargo vessels heading north from Jamaica and Hispaniola invariably passed through TCI's deepwater channels, flanked by secluded cays where buccaneers waited in ambush. More than a thousand ships went to watery graves including one of the most famous wrecks in maritime history—a Spanish treasure galleon called the *Nuestra Señora de Concepcion* that eventually coughed up sixty-five thousand pounds worth of silver, gold, and jewels.

Wherever you venture in Grand Turk, the colony's tiny capital, there is talk of treasure. "I find stuff all the time," said Connie Rus, a Michigan native who runs a local scuba diving outfit. Over the years Connie has amassed a collection of antique bottles: beautiful handblown perfume, medicine, and liquor containers, some still encrusted

with coral. “In the winter, when we get a cold front, the waves get real big,” said Connie. “They break apart coral and grassy areas on the bottom and dump out little treasures.” But those aren’t the only Caribbean destinations that offer aquatic riches. Nearly every isle offers good scuba and snorkeling spots, some of them world renowned: the pristine reefs around Bonaire, the underwater lava flows and black coral of Saba, the coral canyons of St. Croix, and the straits north of Hispaniola where swimming alongside migrating humpback whales is a thrilling experience.

St. Martin, a Windstar destination, is one of the most visited Caribbean cruise ports, boasting an incredible local seafood-based cuisine and a rich mix of French and Dutch cultures.

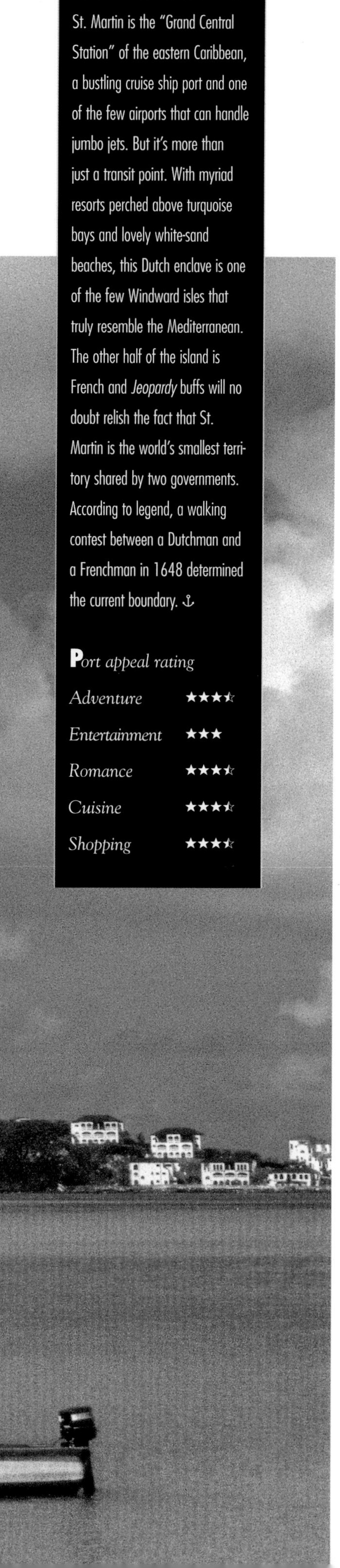

THE SIZZLE OF

ST. MARTIN

St. Martin is the “Grand Central Station” of the eastern Caribbean, a bustling cruise ship port and one of the few airports that can handle jumbo jets. But it’s more than just a transit point. With myriad resorts perched above turquoise bays and lovely white-sand beaches, this Dutch enclave is one of the few Windward isles that truly resemble the Mediterranean. The other half of the island is French and *Jeopardy* buffs will no doubt relish the fact that St. Martin is the world’s smallest territory shared by two governments. According to legend, a walking contest between a Dutchman and a Frenchman in 1648 determined the current boundary. ⚓

Port appeal rating

Adventure	★★★½
Entertainment	★★★
Romance	★★★½
Cuisine	★★★½
Shopping	★★★½

Thoughout the Caribbean, the pleasure is the quest: exploring the region's bays and backwaters, harbors and history, as well as the various myths and legends that make the West Indian islands such fascinating ports of call. One place where you can actually stumble onto real riches—and heaps of treasure stories—is Grand Cayman, one of the region's more popular cruise destinations.

"My favorite treasure story is the one about the German sub," said Dennis Denton, a longtime Cayman resident who's worked several Caribbean treasure expeditions. "It was right after the war and the sub was carrying priceless art pilfered by the Nazis. They were on their way to South America when it hit the reef. There are old guys who swear to me they saw it go down. But no one's ever found it."

One of the more recent yarns is about a honeymoon couple from Atlanta who went snorkeling one morning along Seven Mile Beach. Only a couple hundred yards offshore, they chanced upon the wreck of a sixteenth-century Spanish galleon called the *Santiago*. They returned with scuba equipment, recovered copious gold treasure, and were on the next flight back to the United States.

You don't have to be a scuba diver to explore the natural treasures of the Cayman Wall and its six-thousand-foot drop into liquid oblivion. That's because this is the only place in the world where a "civilian" can hitch a ride on a research submarine. The vessel in question is a petite three-man craft called the *Atlantis* deep-dive submersible. It provides a unique opportunity to see things that were relegated to the realm of Jules Verne until just a few years ago. But it's not for the squeamish nor anyone who is even slightly claustrophobic.

As you start the descent, you get a sinking feeling in the pit of your stomach, the same sensation you get when an airplane hits an air pocket: a blend of gravity force and adrenaline rush. But soon the fear fades away in favor of astonishment as the underwater spectacle unfolds outside the thirty-six-inch-thick hemispherical window. Life is abundant at first, the reef rich with fish and coral that need plentiful light to survive. Then at 220 feet, you back over the edge of the Cayman Wall. It's straight down from there, a black, bottomless pit, and even though you keep telling yourself that they wouldn't let ordinary people do this if it were not safe, the adrenaline starts pumping again, flowing through every vein in your body. The submarine goes into freefall, plunging into the depths of the Caribbean, until there's nothing but dark blue twilight. At about a thousand feet you start to come across the most amazing creatures. Stark white coral shaped like pipe cleaners. Feather stars with long, delicate limbs. West Indian sea lilies, strange umbrella-shaped creatures that have endured on this planet for more than 500 million years. The thrill of observing life-forms that few humans have ever laid eyes upon is unmistakable.

Secluded bays and beaches. Posh resorts and pristine nature areas. Strange quirks of fate. Spirited myths and legends. And some of the most bizarre history you'll ever come across. Indeed, each and every island has its own little treasures. ⚓

Cruise passengers who appreciate native crafts will find a great range to admire and choose from in the Caribbean, including Martinique's dolls dressed in local costumes.

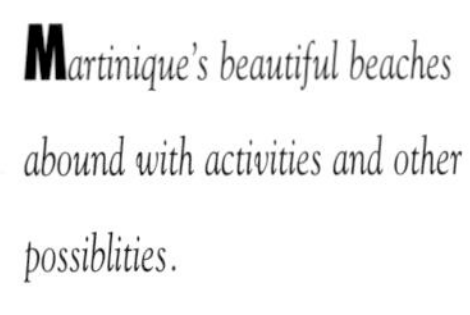

Martinique's beautiful beaches abound with activities and other possiblities.

Aruba boasts more than seven miles of palm-shaded beaches for that perfect romantic stroll at day's end.

THE CARESS OF

ARUBA

Discouraged by its cacti and sand dunes, explorer Amerigo Vespucci shunned Aruba after "discovering" the landfall in 1499. But Vespucci's loss is our gain. Five hundred years later, Aruba is a stunningly different Caribbean island where Dutch gingerbread houses are set against a stark desert landscape. Aruba is one of the region's best sports and recreation destinations. The crystal-clear waters offer an underwater paradise for divers and snorkelers, while the stiff winds that often blow across Aruba offer plenty of power for windsurfers and three-wheeled "land yachts" that speed across the dunes. When all else fails, you can try a bit of the fast action at one of Aruba's ten casinos. ⚓

P*ort appeal rating*

Adventure	★★★
Entertainment	★★★
Romance	★★★½
Cuisine	★★★
Shopping	★★★

Participants share a laugh in Curaçao's annual Tumba, a lively three-day competition to choose the best songs for Carnival.

Overleaf: Caribbean beaches urge a languid bask in the warmth of the sun and a playful splash in the surf.

THE TREATS OF

CURAÇAO

Six islands comprise the Netherlands Antilles, but Curaçao is the one with the most Dutch ambience. The bustling harbor, one of the world's busiest ports, is a tropical Rotterdam fraught with cruise ships, tankers, and freighters bound for the four corners of the globe. Willemstad, the quaint capital, brims with pastel-colored houses reflected in placid canals, like a piece of Amsterdam transported to the Caribbean. Windmills and gabled plantation houses garnish the island's hinterland. This Dutch treat even extends to Curaçao's cuisine, including rijsttafel, a hearty rice dish imported from the Dutch East Indies. ⚓

Port appeal rating

Adventure	★★★½
Entertainment	★★½
Romance	★★★
Cuisine	★★★
Shopping	★★★½

THE PROMISE OF

HAVANA

Havana is a city in transition, a metropolis that hasn't quite found the form that will usher it into the twenty-first century. But that's what makes the Cuban capital so exciting. The old town is still a bastion of Spanish colonial architecture, unfettered by the modern development that has racked so many large cities. The beach areas offer a time trip back to a different era, the flashy 1950s, when Havana was the wildest, naughtiest nightlife town in the entire Caribbean. Mixed in amongst all this bygone spirit are the dour socialist buildings and solemn Soviet-style monuments constructed during four decades of Communism. Havana is all this and more. ⚓

Port appeal rating

Adventure	★★★★
Entertainment	★★★★
Romance	★★★★
Cuisine	★★★½
Shopping	★★½

Havana promises to be an unforgettable cruise destination as it undergoes a rebirth of spirit and hope.

The secret of Guadeloupe's marvelous cuisine is found in the freshness of its daily catch.

THE TASTE OF

GUADELOUPE

Epicures flock to Guadeloupe for some of the Caribbean's best cuisine. Just like Martinique, the island is a full-fledged department of France. And true to their French roots, the island's chefs have elevated eating to a fine art, a savory blend of Creole spices, Gallic sauces, and local seafood. Where else in the Windward isles can you find such delectable dishes as brioche of red snapper with lobster sauce, crayfish salad with foie gras, sea urchin croustillade, or shark gigot with saffron sauce? Many of Guadeloupe's restaurants also have stunning locations, set within old plantation houses and lush tropical gardens. ⚓

Port appeal rating

Adventure	★★★½
Entertainment	★★½
Romance	★★★½
Cuisine	★★★
Shopping	★★½

The exceptionally lovely waters of Buck Island Reef near St. John offer an unparalleled experience for divers and snorkelers.

THE SECLUSION OF

ST. THOMAS

The streets of St. Thomas's Charlotte Amalie boast some of the best duty-free shopping available to cruise travelers. But after hitting your duty-free allowance you might be ready for a change of scenery. A twenty minute ferry ride away is St. John, one of the few Caribbean isles where nature still takes precedence over other leisure pursuits. Roughly two-thirds of the land area is embraced within the confines of Virgin Islands National Park, which sprawls around numerous turquoise bays and over the island's highest point (Camelberg Peak). Park activities include sunbathing, snorkeling, scuba diving, sailing, bird-watching (more than a hundred species), and trekking along twenty-two marked trails. ⚓

Port appeal rating

Adventure	★★★★
Entertainment	★★★
Romance	★★★½
Cuisine	★★★
Shopping	★★★★

The open Atlantic off the beaches of Nevis challenges even the strongest windsurfers.

The scents and colors of Barbados's lush gardens lure cruise ship passengers ashore.

THE BOUQUET OF

BARBADOS

Keeping with good British tradition, Barbados is famous for its parks and gardens, where horticulture is carried out with a flavor and flourish equal to anything you will find back in Mother England. Andromeda Botanic Garden is a fantasyland of tropical plants and flowers perched on a coral ridge above the sea. Welchman Hall Gully in the highland Scotland District offers a nature trail through a limestone gorge planted with herbage from all around the region. Even wilder—but still a carefully crafted garden—is Farley Hill National Park, set around the ruined plantation house where the movie *Island in the Sun* was filmed. Exquisite private gardens are open to scrutiny during the Barbados Horticulture Society's various open-house weekends. ⚓

Port appeal rating

Adventure	★★★½
Entertainment	★★½
Romance	★★★★
Cuisine	★★★
Shopping	★★★

THE CHARISMA OF

ST. CROIX

Perhaps the most underrated of the U.S. Virgin Islands, St. Croix doesn't have the swagger of St. Thomas or the secluded coves of St. John. What you'll find here instead is an enduring sense of history. The narrow lanes of Christiansted and Frederiksted, the only major towns, are flanked by wonderful pastel buildings and stout brick forts erected by the Danish settlers who colonized St. Croix more than two centuries ago. The island also sports some of the best diving and snorkeling spots in the central Caribbean, including the fabulous Buck Island Reef. ⚓

Port appeal rating

Adventure	★★★½
Entertainment	★★½
Romance	★★★½
Cuisine	★★★
Shopping	★★★½

Color abounds throughout the Caribbean culture and its landscape.

Tiny Palm Island in the Grenadines is framed by five white-sand beaches.

Antigua boasts 365 beaches, one for each day of the year.

Overleaf: El Morro fortress, built in the sixteenth century, towers 140 feet above the entrance to San Juan Harbor.

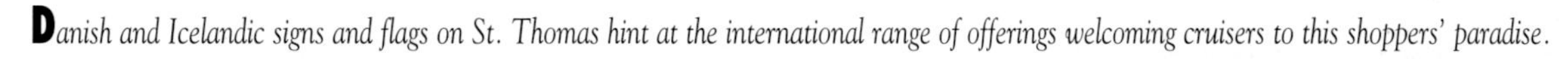

Danish and Icelandic signs and flags on St. Thomas hint at the international range of offerings welcoming cruisers to this shoppers' paradise.

THE TIMELESSNESS OF

SAN JUAN

In certain times and certain places, it seems that San Juan is frozen in time: pacing the walls of El Morro fortress, perched 140 feet above the entrance to San Juan Bay, its bronze cannons still poised for seaborne invasion; sitting in the Church of San Jose, inspecting the Romanesque arches and the thick coral walls; leafing through some of the four thousand historic books harbored within the Casa del Libro; watching a Latin dance recital at the Teatro Tapia; or simply strolling across the Plaza de Armas, where Spanish troops once drilled in the hot tropical sun. There is no more historic port in all the Caribbean, no other city that has done so much to preserve its cherished past. ⚓

Port appeal rating

Adventure	★★★★
Entertainment	★★★★
Romance	★★★★
Cuisine	★★★½
Shopping	★★★½

BY WILLIAM FRIAR

SOUTH AMERICA

LAND OF SUPERLATIVES, LAND OF LIFE

SOUTH AMERICA IS A show-off. The continent shames the meager offerings of less-blessed parts of the planet like an heiress flashing jeweled fingers under the noses of poor relations. It's a land of superlatives. It boasts the world's longest river, the world's highest capital, the world's tallest waterfall. It's a land of extremes. South America has every kind of terrain and climate imaginable, from rain forests to arid deserts, from forbidding mountains to inviting beaches, from cool highlands to steam-bath tropics. And it's a land overflowing with life. Giant Brazil claims one-third of the world's forests. Little Ecuador, about the size of Colorado, shelters more species of birds and plants than the United States.

South America's peoples are nearly as varied, with a history of civilization that dates back thousands of years before a lust for gold and a bad sense of direction steered the Old World to its shores. Though European explorers had been stumbling blindly along its coast for some years earlier, South America's isolation from the rest of the world effectively ended on November 16, 1532. That night the Spanish conquistador Francisco Pizarro began his bloody destruction of the vast Inca empire with a force of fewer than two hundred Spanish soldiers.

Since then the many native peoples of the continent have had to accommodate wave after wave of strange invaders—European colonizers, African slaves, and more recent adventurers, fortune seekers, and exiles from every part of the globe. Five hundred years of clashes and compromise among these peoples have produced some of the richest cultures in the world. Many of these cultures have been confined to the coastline, thanks to the fearsome geography of South America's interior. Steep mountains, dead-zone deserts, and dense jungle still make it difficult even for South Americans to explore their own countries fully. The concentration of coastal life benefits those who want to explore South America by sea. There's an enormous amount to be discovered just by skimming the perimeter of the continent and its many islands.

BUENOS AIRES
MONTEVIDEO
RIO DE JANEIRO
SALVADOR
VALPARAÍSO

It's ironic that a land once famous for mythical lost cities and hidden riches has so many real treasures in plain view along its shores. What follows

Peru's Sacred Valley remains a bastion of traditional Incan customs and costume.

is a brief glimpse of some of the most dazzling of these.

Standing on the massive walls of Cartagena's Old City, visitors might half expect to see a fleet of pirate ships slicing through the Caribbean and heading straight for them. Colonial Cartagena is astonishingly well preserved even though a large modern city has grown up around it. Balconied mansions, deep, dark dungeons, and imposing forts have become museums, restaurants, and shops without losing their old-world charm. The walls that still ring the old city, which is almost completely surrounded by water, are forty feet high and more than fifty feet thick. They are a testament to the riches once kept within and the efforts of freebooters and privateers to pluck them out.

The Spanish founded Cartagena in 1533, and it flourished as one of the most important ports on the Spanish Main. South American gold, African slaves, and European goods flowed into and out of the city for hundreds of years. Though the walls and an elaborate system of forts guarding every approach were supposed to make the city impregnable, it was sacked many times over the years. But its defenses were strong enough to repel a brutal attack by twenty-seven thousand British troops in 1741. Reminders of the city's colorful and occasionally horrific history are everywhere. A magnificent palace contains torture devices used there during the Inquisition. Slave auctions were once held in the largest of the city's many peaceful plazas.

A different kind of living museum can be found on and around the Galápagos Islands, more than a thousand miles to the southwest in the Pacific Ocean. The Galápagos consist of thirteen major islands and dozens of smaller ones peppered along the equator sixty miles from the coast of their mother country, Ecuador. So many places in the world get labeled a "Garden of Eden" that it's easy to become jaded when invited to explore yet another one. And at first the Galápagos actually seem infernal: a dry, barren archipelago forged from volcanic rock. But the creatures who live here convince most visitors this is in fact an earthly paradise. A lack of natural predators has made the wildlife fearless, in some cases even friendly. Sea lions swim right up to snorkelers, ready to play. Penguins, so awkward on land, rocket past underwater, chasing schools of fish. The shore life is so plentiful and so indifferent to humans that visitors have to be careful not to step on an iguana sunning itself on a rock or a blue-footed booby trying to enjoy a quiet moment with its mate. On several of the islands, giant five-hundred-pound tortoises cast a bored eye on passersby, like tired elders who have seen far too much to be impressed by anything.

Santiago's grand plazas flaunt their Spanish heritage.

Visiting Chile by sea, you'd never know it's one of the smaller countries in South America, on average little more than one hundred miles wide. That's because squeezed between the Pacific Ocean and the Andes is a land more than twenty-six hundred miles long, with an endless coastline dotted with port cities. The northernmost of these is Arica, a popular beach resort perched at the edge of the Atacama Desert. This is a region so arid that parts of it have never felt a drop of rain, but on the coast the weather is mild and the water is warm year-round. Landmarks in the town include a nineteenth-century iron church designed by Eiffel, replacing an earlier one swept away by a tidal wave.

Viña del Mar is Chile's premier beach resort.

Halfway down the coast lies Valparaíso, gateway to Chile's fertile and populous heartland. A visitor approaching Valparaíso

Valparaíso charms visiting cruisers with its colorful, cobblestoned, old-world flavor.

THE CHARM OF

VALPARAÍSO

No other South American seaport carries the old-world atmosphere of Valparaíso, a city built by dreamers and schemers who wanted to bring a touch of European elegance to their new home in Chile. Its maze of cobblestone streets and neoclassical buildings are reminiscent of something along the Baltic or Danube, rather than the southeast Pacific. Valparaíso is best explored by foot, especially the tightly packed downtown area. Rickety and romantic old elevators called ascensores, dating from before World War I, help visitors scale the city's notoriously steep hillsides. Nearby Viña, one of South America's most popular beach resorts, also called the Garden City because of its abundance of green spaces and flower beds. The city plays host to Chile's National Botanical Garden, a sprawling expanse of greenery from all around the world. ⚓

Port appeal rating

Adventure	★★★★
Entertainment	★★★
Romance	★★★★
Cuisine	★★★½
Shopping	★★★

A *visit to Machûpicchu, the supreme monument to Incan culture perched high in the Peruvian Andes, is sure to be one of the highlights of a South American adventure.*

from the sea has to absorb the vista in stages. There is the busy port and downtown with its old narrow streets, museums, formal plazas, and the huge national congress building. Behind that jut a crescent of steep hills with a bewildering jumble of buildings, from shacks to great mansions, clinging to their sides. And far off in the distance are the snowy peaks of the Andes. Sixteen ascensores (funicular railways) link the downtown area to the hills. Visitors can lose themselves figuratively and otherwise in the hillside maze of streets, steps, and blind alleys.

Viña del Mar, six miles up the bay, has earned the right to call itself la Ciudad Jardin (the Garden City). Grand mansions preside over carefully landscaped grounds, palms line the streets, and a working clock made from flowers greets visitors as they enter the city. Viña is one of the most famous beach resorts in South America. Museums, the national botanical gardens, a beachside casino, and a raucous nightlife compete with the sea for

One of the world's greatest thrills is to sail around South America's Cape Horn.

the attention of crowds that converge on Viña from throughout Latin America. Every February Viña is overrun with pop and folk musicians of wildly varying talent who compete in the Festival Internacional de la Canción (International Song Festival).

Several hundred miles farther south lies Chile's Lake District. Crystalline great lakes, snowcapped volcanoes, thundering waterfalls, and dense forests put this region high on travelers' lists of the most scenic spots on earth. Anyone familiar with Puget Sound gets a sense of déjà vu on sailing toward the hill-ringed Puerto Montt, which sits at the northern end of a busy gulf studded with islands. Nineteenth-century German immigrants gave the little city an alpine charm, with gingerbread houses adorned with high-peaked roofs and ornate balconies mixed in with plainer ones faced with unpainted wooden shingles. Puerto Montt's frontier-town feel also reminds visitors that this is the last city before the wilds of Patagonia to the south.

The fishing cove of Angelmó, two miles west from downtown,

has waterfront cafés that serve some of the best seafood in a country famous for its fish. Curanto, a seafood stew, is a regional specialty. The crafts market here is also well known as a place to buy handcrafted goods such as boots, copper jewelry, and sweaters made by the Mapuche Indians, a people so fierce they thwarted European settlement of the entire region until late in the nineteenth century.

Nestled at the foot of the Andes, Punta Arenas lies near the southern most point of Patagonia.

A thousand miles of glaciers, mountains, and ancient forests separate Puerto Montt from Punta Arenas, Chile's southernmost city. It clings to a narrow strip of land near the very tip of the continent, wedged between the Andes to the west and the Strait of Magellan to the east. Punta Arenas is nearly inaccessible by land, and approaching it by sea is still an adventure. Hundreds of islands are strewn along the bottom third of the Chilean coast, and between them the sea flows through fjords of breathtaking beauty. Despite its remote location, Punta Arenas is a growing city in a prosperous region. Turn-of-the-century mansions and well-tended plazas are a reminder of its glory days before the opening of the Panama Canal diminished its importance to world shipping.

Magellan, Drake, and Darwin were among the more famous early tourists to sail the waters around Tierra del Fuego (Land of Fire), which got its name from the fires sixteenth century explorers saw burning on its forbidding shores. These were built for warmth by the nearly naked Fuegian Indians, who have long since been wiped out by European disease and persecution. The main island, Isla Grande de Tierra del Fuego, is split between Chile and Argentina. The Argentine city of Ushuaia, on the southern side of the island, is the major port. It claims the title of southernmost city in the world. If not quite the end of the earth, this is as close as many would ever want to come. Towering, icy mountains stop just short of "downtown"; Antarctica is just over the southern horizon. What draws visitors is the region's stark, otherworldly beauty. A national park a few miles west of town boasts glaciers, lakes, rivers, and thick beech forests. Many species of birds congregate here, from steamer ducks to oystercatchers, and there's even a chance of spotting the enormous Andean condor, which has a wingspan of nearly ten feet.

Rounding the tip of the continent and heading north brings visitors to the Atlantic Ocean and the shore of Argentine Patagonia. Inland, this sparsely populated region is dominated by dry, barren expanses interrupted by mammoth glaciers and icy mountain lakes. But its coast teems with life.

At Punta Tombo, about halfway up the coast, hundreds of thousands of Magellanic penguins gather to breed between September and March. A little farther north, the Valdés Peninsula harbors elephant seals, sea lions, penguins, and fur seals. Nearby Isla Pajaro (Bird Island) is home to squadrons of seabirds. Forty-foot right whales often cruise near the coast. Sandy beaches attract a different kind of marine life. This is especially true farther up the coast, which is dotted with seaside resorts like the famous Mar del Plata. In the summer sunbathers blanket its five miles of beaches.

Cape Horn's weather station lures curious tourists.

The Argentine coast ends near Buenos Aires, in every sense one of South America's great cities. Buenos Aires is big, radiating for miles from the edge of the Río de la Plata (Plate River). And it's a cultural as well as national capital, with a world-famous opera house, Teatro Colón, and a festive nightlife that keeps its streets crowded until dawn. Its cafés, wide boulevards, stately gray buildings, and European sophistication remind many of Paris. Several of the city's most prominent buildings are clustered around Plaza de Mayo, just a few blocks from the waterfront. The presidential offices are in the massive Casa Rosada, named for its pink stone facade. North of the plaza, the pedestrian street Calle Florida marks the beginning of the shopping, entertainment, and restaurant districts, where those with enough money can buy a custom-

South America's southern tip is guarded by a stout little lighthouse.

The gaucho of Uruguay is as captivated by the music of the pampas as the Argentinean gaucho.

THE APPEAL OF

MONTEVIDEO

In many respects, Uruguay is a miniature version of Argentina: a facade of European immigrants over a foundation of early Spanish ranchers. Both of these cultures come together in Montevideo, the country's capital, set around the banks of a wide bay and the rolling pampas behind. Soccer, tango, and a good thick steak are still the city's abiding passions. Those with a nose for good shopping will sniff out the city's leather shops where one can order custom-made shoes or browse among handcrafted belts, wallets, and handbags. ⚓

Port appeal rating

Adventure	★★★½
Entertainment	★★★
Romance	★★★½
Cuisine	★★★½
Shopping	★★★

THE MOVES OF

BUENOS AIRES

Argentina's capital is one of the world's great cities, a sprawling metropolis with more than twelve million people on the banks of the Río de la Plata. It's a huge melting pot with an eclectic population derived from Italian, Spanish, German, and Slavic immigrants. This Latin Manhattan never seems to stop, the downtown streets and sidewalk cafés are filled with people until way past midnight. The city's nightclubs swing to the beat of tango, that most romantic of all dance forms, while the popular parrilla restaurants specialize in another Argentine mainstay—beef roasted over an open fire. ⚓

Port appeal rating

Adventure	★★★★½
Entertainment	★★★★½
Romance	★★★★
Cuisine	★★★★
Shopping	★★★★

made leather jacket, sit down to a brick-size Argentine steak, and take in a play within the space of a few blocks.

Buenos Aires is a city of barrios, or neighborhoods. The old port of La Boca has a dilapidated quaintness, its parti-colored houses resembling harlequins down on their luck. To its north is San Telmo, the artists' quarter. It has some of the city's oldest architecture and many places to experience the tango. The tiny residential neighborhood of La Recoleta is best known for its exclusive cemetery, a true necropolis where the tombs are grander than many people's homes. Eva Perón is buried here.

Iguassu Falls on the Iguaçu River divides Paraguay, Argentina, and Brazil.

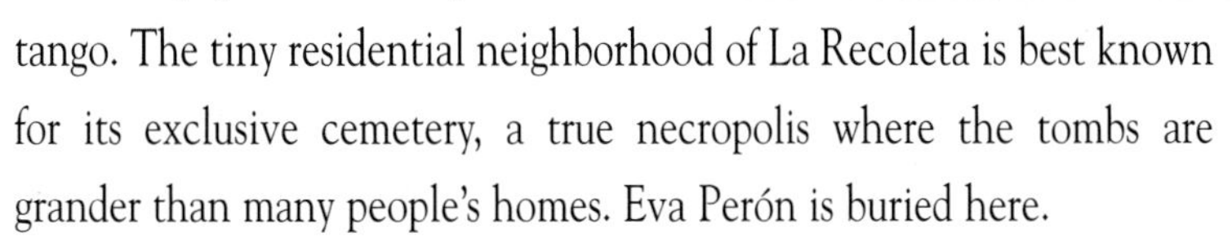

The Río de la Plata is not actually a river but an enormous estuary where the Paraná and Uruguay rivers meet the Atlantic. On its eastern shore lies Montevideo, the capital of Uruguay and its major city. Though just a few miles east of Buenos Aires, Montevideo has a much calmer, small-town feel. This is not from lack of sophistication; Uruguay is one of the most highly educated countries in South America, with more than its share of writers and artists. The exquisite acoustics of Montevideo's grand Teatro Solís draw musicians, dancers, and actors from around the world. One of the city's most popular museums is the Museo de Gaucho y de la Moneda (Cowboy and Coin Museum), which documents the history of the gaucho, as fabled a part of Uruguay's history as Argentina's.

Strolling musicians lend atmosphere to the Montevideo waterfront.

The sandy beaches at the edge of the city become increasingly beautiful and unspoiled as one heads east along the Uruguayan riviera. Exclusive Punta del Este is the most famous of many coastal resorts. The beaches continue for hundreds more miles up the coast of Brazil, the fifth largest country in the world. Brazil takes up half the land mass of South America, but for many visitors it means one place: Rio de Janeiro.

Rio, like many of its denizens, is drop-dead gorgeous. Deep blue waters break along its fifteen-mile coast, which is backed by mountains covered with emerald tropical forest. Craggy granite peaks jut up throughout the city, as though deliberately giving visitors lookout points to admire its beauty. The best known of these is Pão de Açúcar (Sugar Loaf), which offers heart-stopping views from its summit. To the left is Guanabara Bay and Rio's frenetic downtown. To the right is the vast sandy half-moon of Copacabana, the first in a string of beaches that seem to go on forever. Turning inland, one sees Corcovado (Hunchback Mountain), rising nearly a half mile above the city. The famous one-hundred-foot statue of Christ the Redeemer stands at its peak, literally greeting visitors with open arms.

Not everyone is so welcoming. Rio has become as notorious for crime as it is celebrated for good times. There is plenty of suffering, too. Desperately poor people crowd hillside favelas (shantytowns), looking down on the wealth of the city from shacks with million-dollar views.

But Rio's Carnival remains the most festive costume party in the world, a marathon spectacle that takes over the entire city in the week before Ash Wednesday. It's serious fun: rival samba schools begin rehearsing and planning costumes months beforehand.

As if Rio weren't plenty for any one country, Brazil also has Salvador da Bahia, nearly a thousand miles up the coast. Like Rio, its location is physically stunning. Salvador is perched at the tip of a peninsula overlooking the sweeping Bahia de Todos os Santos (All Saints' Bay), at four hundred square miles the largest bay on the Brazilian

The Argentine flag was the inspiration for this gaucho spur.

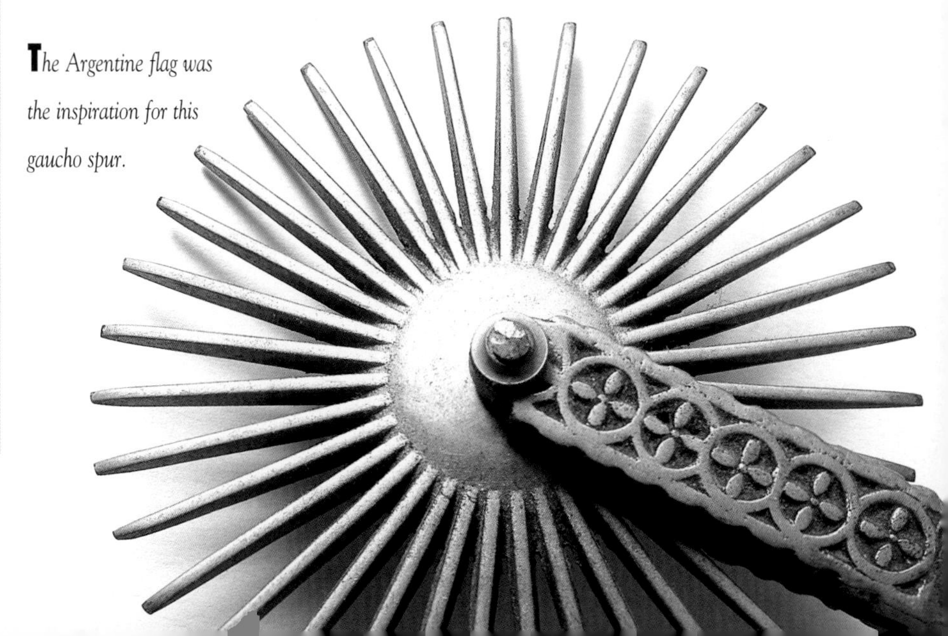

Tango's recent revival mirrors a cultural renaissance throughout Argentina.

THE VISTAS OF

RIO DE JANEIRO

Rio is the kind of place that steals your heart and takes your breath away. Set around the shores of sparkling Guanabara Bay, backed by jungle mountains and flanked by palm-shaded beaches, it would be difficult to find a more beautiful city. Copacabana and Ipanema are a real treat for the eye, not just the fine white sand, but the abundance of bronzed bodies—both male and female—that strut up and down the strand. Corcovado, with its sweeping views and majestic statue of Christ, is one of the world's most famous peaks. And if you happen to come for Carnival, the city transforms into an endless conga line of raunchiness and revelry. ⚓

Port appeal rating

Adventure	★★★★½
Entertainment	★★★★½
Romance	★★★★½
Cuisine	★★★★
Shopping	★★★½

Sunrise silhouettes the myriad hills and bays of Rio de Janeiro.

coast. Salvador has miles of spectacular beaches, but it offers much more than that. The upper part of the central city, Cidade Alta, sits atop a cliff and preserves centuries-old ornate churches, cobblestone streets, and brightly painted colonial homes. A huge elevator connects Cidade Alta with the busy streets of Cidade Baixa (Lower City), the heart of the business district.

Salvador's rich cultural life is highly Africanized. The city has more than 150 Catholic churches, but it has many times that many terreiros (temples) dedicated to Candomblé, a colorful Afro-Brazilian religion. Capoeira, a martial art disguised as a dance, is still performed. African slaves once practiced it under the noses of their Portuguese masters. The food here is exotic. The staple dish is moqueca, a delicious seafood stew cooked in a clay pot with coconut milk, palm oil, and peppers. Equally exotic are the many kinds of music, dance, and art that seem to spill out of every corner of the city. Salvador is a uniquely festive place. Carnival here rivals that celebrated in Rio.

Sugar Loaf overlooks the entrance to Rio's harbor.

Most of these ports are destinations in themselves, and there are many more along the shores of South America. But even the grandest port city doesn't begin to hint at what lies inland. Often the people, wildlife, land, and the climate of the interior have little in common with those found on the coast. No wonder so many voyagers have returned again and again to South America, determined to explore it a little more fully. One does not even have to leave ship to feel the pull of the continent. Standing on deck at night, one can squint past the harbor lights, close one's ears to the blare of the city, and sense the presence of a new world still waiting to be discovered. ⚓

Even her souvenirs reflect the variety and ingenuity of Rio's people and culture.

Corcovado and its famous Christ statue endure as global symbols of Rio.

Rio's annual Carnival retains all of its famed splendor and splash.

Some of Salvador's color derives from the beat of its Afro-Brazilian music.

Overleaf: *The Amazon Basin astonishes travelers with the sheer expanse of its jungle and waters.*

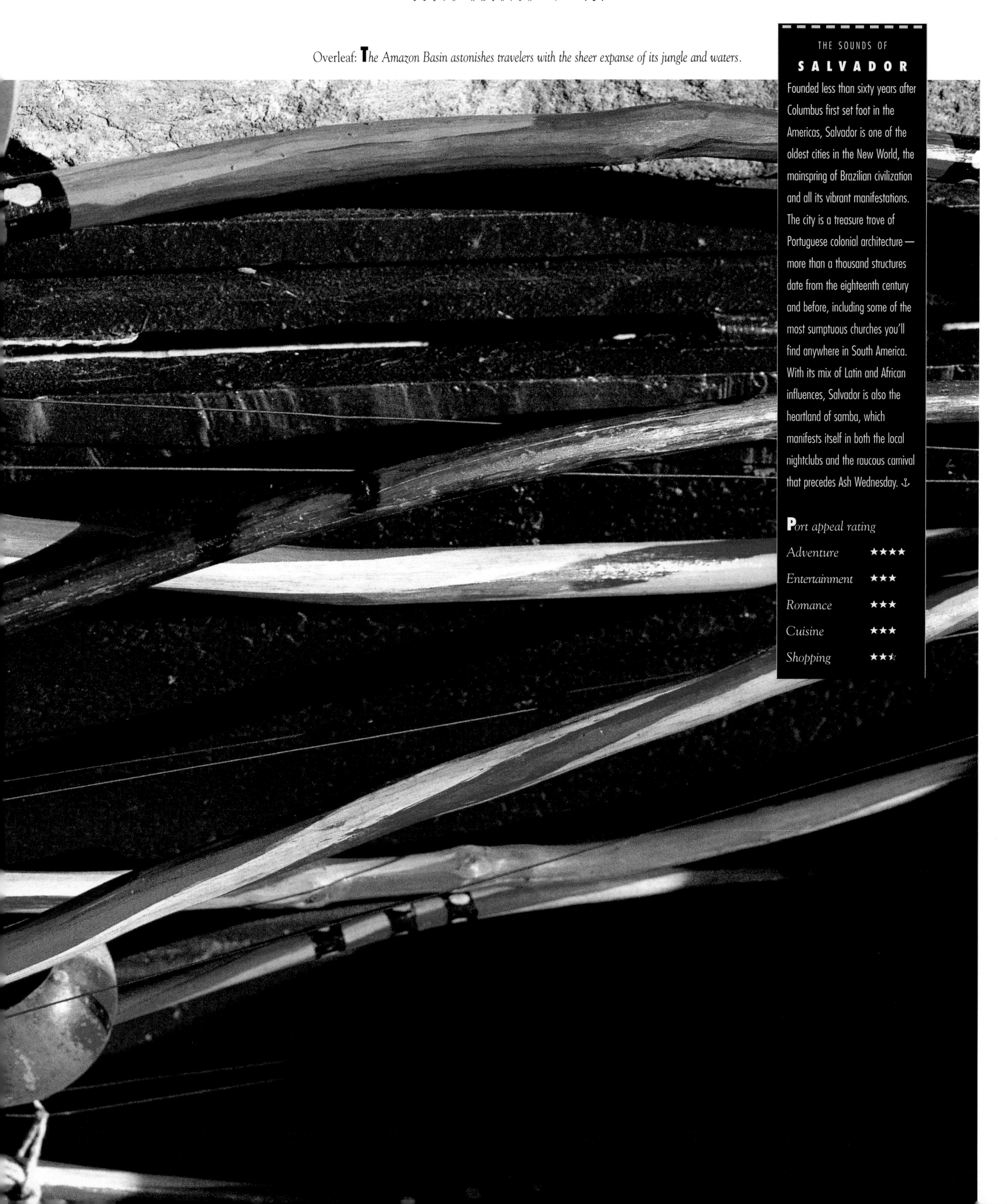

THE SOUNDS OF

SALVADOR

Founded less than sixty years after Columbus first set foot in the Americas, Salvador is one of the oldest cities in the New World, the mainspring of Brazilian civilization and all its vibrant manifestations. The city is a treasure trove of Portuguese colonial architecture — more than a thousand structures date from the eighteenth century and before, including some of the most sumptuous churches you'll find anywhere in South America. With its mix of Latin and African influences, Salvador is also the heartland of samba, which manifests itself in both the local nightclubs and the raucous carnival that precedes Ash Wednesday. ⚓

Port appeal rating

Adventure	★★★★
Entertainment	★★★
Romance	★★★
Cuisine	★★★
Shopping	★★½

BY CHRISTOPHER P. BAKER

C·E·N·T·R·A·L A·M·E·R·I·C·A

CRUISING THE OLD SPANISH MAIN

ACAPULCO . . . PANAMA . . . COZUMEL . . . Places steeped in the allure of the old Spanish Main. Imagine the sense of discovery as you set out in search of the ghosts of lost civilizations, retracing the route of Spanish galleons with stops at exotic ports that once served as conduits for the wealth of an empire. Just the written itineraries seem to give off the scent of hibiscus, the sounds of merengue and mariachi serenades, and—close your eyes—the sensual caress of a hot sun on soft sandlike talcum dissolving into waters the color of liquid gems.

To be borne to these pearls on a modern cruise ship only adds to the drama.

The ports of call are all in the reliably warm and romantic tropics that transform the routine of a drab northern winter into the joys of June. In other regards, they're as varied as they are numerous, sprinkled like far-flung pirate's treasure along the Pacific and Caribbean coasts of the isthmus. Some have changed quite a bit with the years: Acapulco still has its cathedral and castle, but what lures is the shopping, plenty of beer and tequila, and an array of activities to dazzle the experienced or intrepid cruiser. Others, such as Cancún and Zihuatanejo, weren't even on the maps of the conquistadores. Nor did Hernán Cortés get to thread through the narrow Panama Canal—one of the world's great cruising thrills.

There are two sides beyond those of east and west. The first is the jet-set world of the Mexican Riviera, exemplified by Acapulco and Ixtapa, with their diamond-dust beaches, rum-swizzle excursions, discos, water sports, and bargain shopping. The other side of the coin, the one you don't hear much about, is the realm of offbeat adventure offered by rarely visited ports such as Puerto Caldera, Puerto Cortés, and Puerto Cristóbal: from climbing jungle-clad Mayan ruins to unraveling the mysteries of the tropical rain forest while white-water rafting or moving languidly through the treetops of Costa Rica aboard an aerial tram.

ACAPULCO
CABO SAN LUCAS
IXTAPA
MAZATLÁN
PLAYA DEL CARMEN
PUERTO VALLARTA

Many cruisers think of Mexico and Central America as a single destination and know little of

Lucky cruisers to Panama's San Blas Islands enjoy the artistry of the Cuna indians reflected in their bright colors and traditional motifs.

Gathering in her visitors and residents alike, Acapulco radiates the warmth of the Mexican people and culture.

the diversity of the individual nations. Though these countries have some things in common, each has its own character and distinct culture, with several layers of life to discover for the inquisitive traveler.

Mexico, for example, changes its mask as frequently as the dramatis personae of a Cervantes play. From jungle to desert, Mexico has a mood for all seasons. Take Cancún and Cabo San Lucas, both boasting snowy beaches and peacock blue seas. But the former is a sophisticated and lively resort of luxury high-rise hotels framed by the emerald jungle of the Yucatán, while the latter is a lackadaisical place of repose backed by pastel desert shimmering in the heat like a sort of dreamworld between hallucination and reality at the tip of the slender Baja California Peninsula.

"Cabo," a favorite among West Coast cruisers, is a cornucopia for pursuits as diverse as luxuriating on secluded Playa del Amor, hiking the rugged Sierra mountains, or making the most of King Neptune's bounty . . . and not just the seafood. For eyeball-to-eyeball encounters with whales, the warm waters off Baja's sunny Cape are unrivaled. Thar she blows! There is no greater epiphany at sea than leaning over the rail to marvel at Cabo's great whale parade. And when cruisers tell you about the big one—the really big one—that got away, don't believe it. Cabo is a sportfishing paradise. Heck! Name a Mexican port that isn't. Hard-fighting marlin swim in these waters year-round, and seem to be lining up to get a bite on your hook.

Mexico's quintessential port city—the yang to sleepy Cabo's yin—is surely Acapulco, once the greatest pleasure resort in the world. Thirty years ago there was no place quite like it for lazy siesta days and sensuous tropical nights. Well-heeled cognoscenti flocked in droves, lured by an awesomely beautiful bay, winter weather as sublime as anywhere in the northern hemisphere, and a salsa caliente nightlife compared to which all others seem faint-hearted efforts. Then Mexico's most illustrious resort lost its sheen. It had become vulgar. Overcommercialized and overdeveloped. After a major remake, the "Queen of the Pacific" is back, more glamorous than ever and in a way unlike any other Mexican resort. If you haven't been for a while, the difference is night and day. Literally. Noisy, brash, fun-loving Acapulco never sleeps!

The great variety of Mexican handicrafts available to travelers includes pottery and intricately carved masks.

Everything you could possibly want in a resort port is here. Exquisite beach hotels ranging from the picturesque to the palatial. Championship golf courses. World-class tennis. Water sports such as parasailing (easier and safer than it looks, and every bit as much fun). Restaurants that have remained through the years sentimental favorites of fashionable café society. And fabulous shopping, be it in ritzy boutiques selling brand-name jewelry and resort wear; galleries offering traditional and contemporary art; the bustling markets of the old town, where Acapulqueños tout exotic homespun fabrics and crafts; or even Taxco, nestling amid verdurous mountains and the place to bargain for its world-renowned silver jewelry. Taxco is itself a crown jewel, to be enjoyed while walking its narrow, winding lanes and cobbled plazas studded with baroque monuments from its silver-rich past.

Fearless divers can take the plunge at Acapulco's Quebrada cliffs.

Acapulco's famous high-divers still perform their spectacular leaps, plummeting over one hundred feet into the ocean—like diving out of a fifteen-story window. And the mariachis and margaritas are always superb, giving foretastes of Mexico's most exciting vida nocturna. The city doesn't hit its stride until the stars reach their zenith, when Acapulqueños rouse from the day's torpor and head down to the shorefront Costera, now a zany, glittering river of neon and pulsating discos . . . perfect for kicking up a little dust before reboarding. Bidding good-bye never comes easy.

Between Acapulco and Cabo are a string of ports, evenly spaced, as if conceived for leisurely cruising. Along the shore, the region's beaches rival in beauty and solitude those of anywhere in the world. A computer couldn't have figured it better. Or could it? The Mexican Riviera has expanded in recent years to adopt the sleepy little fishing

Some of Mexico's finest beaches wrap themselves around Acapulco Bay.

THE ENCHANTMENT OF

ACAPULCO

Movie and music stars have flocked to Acapulco since the 1920s when it first blossomed into a world-class resort. One of the first celebrities to discover its steamy charms was swashbuckling Errol Flynn, but over the years this swank Mexican port has attracted a number of regular denizens, including Johnny Weismuller and Elvis Presley. One of the main attractions is Acapulco's rowdy nightlife. The city sports more than 120 discos ranging from beachside dives where bathing suits are de rigueur to posh hotel dance halls with strict dress codes. Other after-dark charms include flamenco clubs, folklorico shows, or merely walking beneath the stars on one of the wonderful white-sand beaches. ⚓

Port appeal rating

Adventure	★★★½
Entertainment	★★★★
Romance	★★★★
Cuisine	★★★½
Shopping	★★★½

THE ALLURE OF

PUERTO VALLARTA

The pace was set by Richard Burton and Liz Taylor, a steamy liaison in Puerto Vallarta during the filming of *Night of the Iguana* in the 1960s. Now this Mexican Riviera destination is one of the most romantic places in all of Latin America. The setting is absolutely stunning: white-sand beaches framed by the jungle-shrouded Sierra Madre and the sparkling Bahia de Banderas. There are several ways to enchant your lover: a sunset walk along the beach, candlelight dinner in one of the many outdoor cafés, or perhaps a champagne flight in a hot-air balloon. ⚓

Port appeal rating

Adventure	★★★½
Entertainment	★★★
Romance	★★★★
Cuisine	★★★½
Shopping	★★★½

village of Zihuatanejo and its nearby "billion-dollar beach" at Ixtapa, chosen by computer—as was Cancún—to be metamorphosed from scintillating sand and sea into a glittering resort for chic cognoscenti. The old and the new, the languorous and the luxurious, side by side. Both small and easy to know. What a treat! It's enough to make you and your significant other want to jump ship . . . drink a toast of tequila . . . slip out of your clothes . . . take a refreshing swim in aqua blue waters . . . kiss and make love . . . and spend the rest of the week getting drunk on the romance of it all.

Looking for colonial quaint? Try Puerto Vallarta, or PV for short. This charmer was an obscure fishing village until 1963 and the filming of *Night of the Iguana*, whose two stars, Elizabeth Taylor and Richard Burton, fell in love with the town as much as each other, bringing PV a celebrated romance and glamorous fame. Mexico's third most popular tourist resort has maintained a barefoot, casual charm typified by the colorful old neighborhood known as El Centro, with cathedrals and plazas redolent of former colonial splendor. Basket-laden burros still roam cobblestone streets lined with red-tile-roofed, whitewashed houses with bougainvillea and hibiscus spilling giddily over their walls. It's hard to imagine a more conducive ambience for romance to bloom.

Playa la Ropa beach is just one of Zihuatanejo's many breathtaking beaches awaiting cruisers.

By contrast, Mazatlán, northernmost of Mexico's mainland Pacific resorts, is a working port city, earthy, and one that doesn't rely on the tourist trade. Young people have a special fondness for the "Pearl of the Pacific," where the style is carefree and casual: it's perhaps the most laid-back of the Pacific resorts. In the old quarter, where life goes on in the traditional Mexican style, mariachi bands may strike up a tune in the parking lot just for the fun of it. Like every resort port, there are beaches—a seventeen-mile-long sweep of creamy sands—and seafood, fresh and succulent, straight from the briny blue. Of course! The city sits at the mouth of the Sea of Cortez where it joins the Pacific Ocean—a pelagic playpen for schools of marlin, dolphin fish, wahoo, and tuna, landed daily, still wriggling, onto the waterfront docks. Mazatlán is the place for big game fishing . . . its biggest lure.

Puerto Vallarta boasts twenty-five miles of beachfront at the foot of jungle mountains.

Mexico's Caribbean shores offer no less enticing cures for the midwinter blahs. Cancún has skyrocketed from the sands to become an in-vogue temple of ritual narcissism, while more adventurous souls can discover their own beauty inland, rummaging the jungle-framed Mayan temples of Chichén Itzá and Tulúm. Cozumel, a sand-trimmed island off the Yucatán, is another cruise-ship staple, often combined with a stop at Playa del Carmen, a somnolent fishing village closer to the temple ruins. Bring your flippers! These waters, limpid as liquid light, vibrate with color and life, enticing snorkelers and divers to discover a world more beautiful than a casket of gems.

Masks are indispensable to many Mexican fiestas.

Cozumel. Ixtapa. Mazatlán. Treasure troves, one and all, for native shopping, be it for avant-garde art; black Oaxaca pottery; carved onyx animals; the distinctive beadwork, masks, and jewelry of the Huichol Indians; a terrific pair of Mexican huaraches (handmade sandals); or even tapestries woven by Indian women on traditional backstrap looms. And wherever you come ashore, Mexico offers heaps of streetside food stalls (taquerias) and palapa-covered restaurants and bars perfect for a long, lingering lunch alfresco, where the drinks are strong and you can savor grilled fish, lobster, or regional specialties with your feet practically in the water.

Who hasn't dreamed of cruising from sea to shining sea through one of the great man-made wonders of the world? The fifty-mile transit of the Panama Canal between Colón on the Caribbean side and Balboa on the Pacific is a nine-hour passage invariably made in daylight as the centerpiece of ten-to-twenty-one-day, one-way cruises. Ships stairstep the continental divide by a series of six massive locks, sail across gargantuan man-made Lake Gatún, and slice through the narrow, eight-mile-long Gaillard Cut past forested emerald landscapes within fingertip reach.

Usually passengers stay aboard. Other ships, such as World Explorer's *Universe Explorer*, don't even enter the "big ditch." Instead, they tie up at Colón and use it as a base for exploring a country that is much more than a canal between seas.

Panama stirs visions of "gold coast" glory with excursions to erstwhile booty-filled cities such as Portobelo, founded in 1597, and

Fine Mayan stone relief carvings lure visitors to the famed Chichén Itzá.

THE TALES OF

IXTAPA

Call it a tale of two cities. Zihuatanejo is a traditional old fishing village on the Pacific coast. Ixtapa is the modern, created resort a few miles away. The two live in a sort of shotgun symbiosis, Zihuatanejo having lost some of its innocent beauty to tourism while Ixtapa gained a picturesque counterpart. Ixtapa's beach is lined with high-rise hotels; Zih's has small hotels tucked under palm trees or hugging rocky cliffs. Ixtapa is glitzy; Zih clings to its old ways. Somehow it works. Each city gains something from the other. ⚓

Port appeal rating

Adventure	★★★½
Entertainment	★★★
Romance	★★★★
Cuisine	★★★½
Shopping	★★★

Deserted beaches reflect the natural beauty and restorative quiet offered to cruisers by Zihuatanejo.

The beautiful resorts of Mazatlán promise lounging beside man-made swimming pools and frolicking on white-sand beaches.

THE PLEASURES OF

MAZATLÁN

Most people come to Mazatlán for the longest beach in Mexico, sixteen miles of unbroken sand between Cerro del Creston and Punta Sabalo. But this sprawling northern port has another claim to fame: it's the only Mexican beach resort that has managed to preserve much of its colonial heritage. Founded more than 450 years ago, much of the area around Plazuela Machado dates from the eighteenth century and is currently being restored. Its architectural gems include the Basilica of the Immaculate Conception, the Angela Peralta Theatre, and the Mercado Suarez with its lively market. ⚓

Port appeal rating

Adventure	★★★½
Entertainment	★★★
Romance	★★★½
Cuisine	★★★½
Shopping	★★★

Panama City, fast-paced and exuberant, with antique quarters—their narrow streets festooned by filigreed balconies—standing in the shadow of skyscraping banks and hotels and casinos. And flight excursions can whisk you out to the San Blas Islands, inhabited by Cuna Indians known for their exquisite molas, traditional appliquéd panels.

The darling of eco-tourists is neighboring Costa Rica, where nature takes front stage and wildlife loves to put on a song and dance. No wonder adventure cruise ships are flocking here like migrating macaws. There's more to this peaceful nation than shooting wildlife through the lens of a camera. Golfers, scuba divers, surfers, white-water rafters. Every kind of active cruiser can find nirvana in Costa Rica: about the only activities not offered are those that involve snow skis or camels. It's a medley befitting a diminutive country that is really a microcontinent. Anyone who wants to journey, as it were, from the Amazon to a Swiss alpine forest has simply to start on the coast and walk uphill. Within minutes of Puerto Caldera, on the

Pacific, the tableau changes with dramatic effect from dry savanna to mist-shrouded cloud forest swathing the slopes of volcanoes spiced with steaming hot springs and lakes boiling with guapote (rainbow bass).

A favored destination is the world-famous Monteverde Cloud Forest Reserve, where the air is cool and dank and alive with the mournful whistle of the quetzal—the Holy Grail of tropical birds. Puerto Limón, on the Caribbean coast, offers Tortuguero National Park,

At the southern tip of Baja California, land comes to a spectacular end in Cabo San Lucas.

THE FLAVOR OF

CABO SAN LUCAS

If Hemingway were alive today, Cabo San Lucas would probably be one of his regular haunts. Not so much the rowdy bars and pristine beaches, but the harbor area where sportfishing boats take to the sea each day in search of marlin, swordfish, sailfish, and other "big game" species. Marlin can easily run more than a thousand pounds, a test of both your strength and wit. Most people "catch and release" these wonderful fish, opting to capture their moment of triumph on film or video rather than with a trophy mounted on the living room wall. There are plenty of other fish to fry, dorado and wahoo in particular. ⚓

Port appeal rating

Adventure	★★★★
Entertainment	★★★
Romance	★★★★
Cuisine	★★★
Shopping	★★★

a world of close-canopied rain forest as sweet-smelling and innocent as it must have been in the first light of Creation. You feel like Lucy entering the magical kingdom of Narnia in C. S. Lewis's *The Lion, The Witch, and The Wardrobe*.

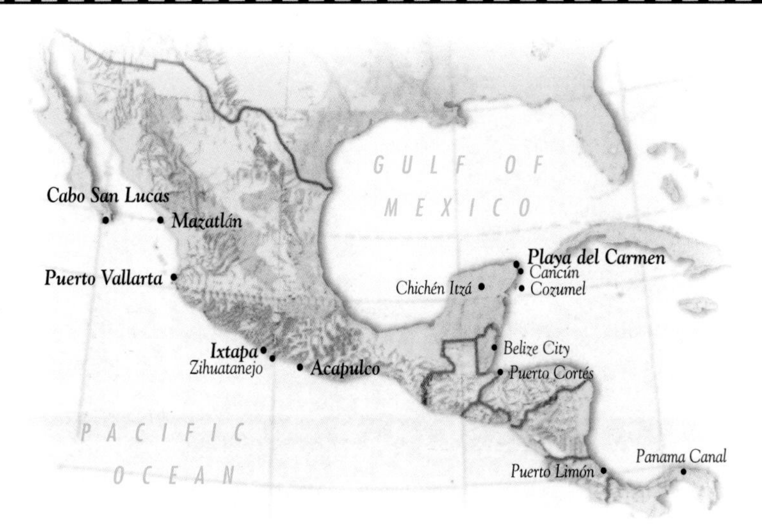

Oh, such sounds and colors! Toucans like rainbows; hummingbirds like the green flash of sunset; poison-arrow frogs with liveries brilliant enough to scare even the most dim-witted predators; and electric-blue morphos, the neon narcissi of the butterfly world. Every few yards seems to bring something new. Jaguars. Monkeys. Sloths like smiling Muppets moving languidly among the high branches. Stay here a while and you'll begin to think that you might, like Noah, see all the creatures on Earth. Excursions are usually civilized, but with an experienced guide—hacking away a vine now and then with a machete—to add just the right jungle mystique.

Though the history buff may be disappointed by Costa Rica's lack of pre-Columbian or colonial sites, Belize, Mexico's Yucatán, and Guatemala—especially Guatemala—make amends. Tikal, the most spectacular city of the Mayan civilization, is enough to bring out the Indiana Jones in any cruiser, bursting up from lowland jungle that is in the same pristine condition as it was when the ancient Maya lived under its green canopy. And Antigua, the former Spanish capital of eighty-thousand people, is pickled in aspic as it was in 1776 when its life was snuffed out by an earthquake. The splendor of the past permeates through the soles of your feet as you walk the cobblestoned streets, marveling at its setting of great natural beauty against a backdrop of looming volcanoes. These and other treasures lie within an hour or so's drive of Guatemala's Puerto Quetzál.

As if that weren't enough, Guatemala is a trove of ethnic items: exquisite jade jewelry and Indian handwoven textiles, including huipiles, the beautiful blouses worn by highland women. Even Costa Rica has its artisans' center. Sarchí is one of Central America's foremost crossroads of crafts, nestled dramatically in the Central Highlands amid row upon row of coffee bushes, dark and shiny, cascading downhill like folds of green silk. Sarchí is most famous for the intricately detailed, hand-painted oxcarts that originated here last century and which can still be seen being pulled along dusty back roads by plodding bovines led by leather-skinned old men.

Reclining with a cocktail on the sundeck as your ship sails away, you ruminate on the fireflies of your recollection, as real fireflies flash by, and the silhouettes of great galleons, slipping out to sea, laden with treasure en route to Spain. ⚓

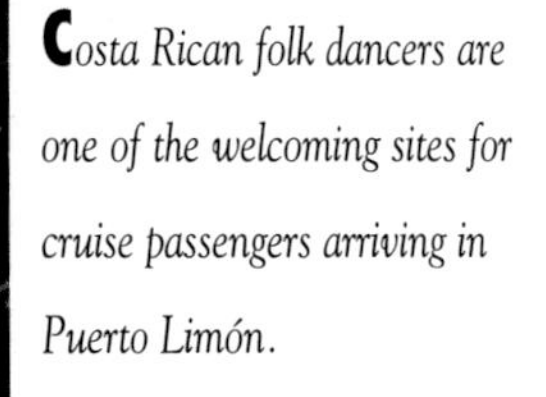

Costa Rican folk dancers are one of the welcoming sites for cruise passengers arriving in Puerto Limón.

Architectural treasures dot the old colonial towns of highland Costa Rica.

Right: Passengers on the Royal Viking Sun *get an up-close view as they pass through Gaillard Cut, the narrowest part of the Panama Canal.*

ROYAL VIKING SUN

THE ANTIQUITIES OF

PLAYA DEL CARMEN

Perched near the tip of the Yucatán Peninsula, Playa del Carmen is a gateway to the many seaside attractions of Mexico's Caribbean coast as well as the region's archaeological relics. The ruined city of Chichén Itzá, the most famous of all Mayan sights, lies amid the jungle about a three-hour drive from Playa del Carmen. Many of the city's eighteen temples and palaces have been restored. The ruins of Tulum, dominated by the grand palace and the castle, overlook the turquoise Caribbean about forty-five minutes from Playa del Carmen. ⚓

Port appeal rating

Adventure	★★★★
Entertainment	★★★
Romance	★★★½
Cuisine	★★★
Shopping	★★★

El Castillo towers over the ancient Maya ruins at Chichén Itzá.

BY JO BETH MCDANIEL

WEST COAST

THE WILD AMERICAN PACIFIC RIM

A CLASH OF TITANS sculpted the West Coast: the mighty Pacific Ocean crashing against glaciers, which in turn were carving out the jagged peaks left by strands of volcanoes and the grinding of continental plates. The result is eight thousand miles of rugged, unique, and uncommon beauty, from San Diego north to Anchorage.

Here, the fertile coastal valleys are framed by mountain ranges, lined up to the east like a fortress against a more ordinary world. These snow-covered peaks protected this shore from European-style civilization until late in the eighteenth century, when a few hardy souls founded settlements at San Diego and Los Angeles.

After that, it didn't take long for word to spread of the riches here: the enormous pods of whales and porpoises, the colonies of sea lions stretched out along the rocky shoals, the cool summers and mild winters that allowed local natives to live virtually naked throughout the seasons.

Today, San Diego and Los Angeles are two of the largest, most vibrant cities in the world, famous for their sunny beaches and laid-back lifestyles. San Francisco has bloomed into an unusually enchanting city, as has Seattle, both of which are known as much for natural beauty as for their local creative geniuses. Vancouver, tucked onto a gorgeous water-wrapped peninsula, attracts so many thousands of new residents that its skyline changes dramatically from year to year. Far to the north, Alaska is experiencing another gold rush, this time wrought by cruising tourists who flock here to experience its unmatched wilderness.

So much has changed, yet so much is, remarkably, the same. The city hillsides may be covered with luxury homes, but just beyond them, one can still see the same rugged coast that enthralled those early explorers. The waters still teem with sea life, and many forests remain as wild and inaccessible as they were in centuries past. Today, locals wear a few more clothes, but of course, the weather is still wonderful.

Ocean currents command these shores, tempering summer's heat and winter's wrath. Sun-drenched

LOS ANGELES
SAN DIEGO
SAN FRANCISCO
SEATTLE
VANCOUVER

Sitka showcases one of Alaska's finest collections of native artifacts including totem poles.

San Diego, at the desert's edge, remains cool and pleasant even in summer, while at the north end of the region, along Alaska's Pacific coast, winters tend to be milder than those in the Midwest.

The ocean's fogs nurture the land, coaxing growth from plants seen nowhere else. Here stand towering redwoods and massive sequoias, the largest living things on earth, alongside bristlecone pines, the oldest living things on earth.

Flowers luxuriate in the mild weather. Gardens burst forth with unusual plants from Africa and Europe, Asia and Latin America, all blooming beautifully together. Here, too, are people from all over the world, drawn as if by a beacon, seeking a place in this golden land of opportunity. The cities hum with their diversity, the vast array of ethnic celebrations and cuisines, and rich cultural traditions.

Fittingly, there is not just one landscape along this coast, but a dozen—from the continent's highest mountains to its lowest desert valleys, and deep, dense forests filled with bears and mountain lions, elk and deer. In any West Coast city, even L.A., you are never far from a deserted beach or forested wilderness, if you know where to look.

In San Diego, the Pacific beckons at every turn. Downtown is set smack on the bay, so the entire city seems like an extension of the ocean. The city's watery edge changes mile to mile, from surf-friendly beaches to the elegant coves of La Jolla and marinas rustling with white sails. Everywhere there are flowers, from dramatic stands of honeysuckle to the magenta flash of bougainvillea vines cascading down crisp white stucco walls.

The city stretches out at the border of Mexico, and draws much of its charisma from its Spanish roots. In Old Town, a leafy park with historic adobes, the air is thick with the aroma of handmade tortillas sizzling on the grill. The city snaps to the captivating rhythms of its roaming mariachi bands; the celebratory colors of piñatas and Mexican blankets decorate many a wall.

San Diego's most enchanting refuge is Balboa Park, home of the world-famous zoo. Stately buildings, left over from the 1915 International Exposition, house museums and theaters. The buildings are so ornate, the landscaping so lush, that it feels like a bit of Florence transported to the tropics.

Downtown, the Gaslamp Quarter's lively old avenues have been reclaimed by jazz bars and outdoor cafés overflowing with chic young crowds. Another treasure is just a ferry ride away, on Coronado Island. Here, the Hotel Del Coronado, built in 1888, perches like a regal queen on its own tranquil beach.This national historic landmark is a wonderland of wooden whimsy, an elegant time capsule back to the Victorian era.

On a clear day—and when is it not sunny and beautiful in San Diego?—climb to one of the high bluffs along the Pacific for a panoramic view of the shimmering blue expanse. During the winter, Point Loma is best for spotting whales headed south to the calm lagoons of Mexico. At Torrey Pines State Reserve, you can sit under the rare, wispy pines and watch dolphins surfing the waves that break on the beach below.

A hundred miles to the north, past a string of funky little surf towns, sits Los Angeles, an urban sprawl larger than some Eastern states. The city is a constant parade—a drama, comedy, and magic show rolled into one. Even the snowcapped mountains to the east seem controlled by a sorcerer's hand: invisible one minute, stunningly close the next. Here, just beyond the city, are dense forests, hidden waterfalls, pure alpine lakes, and, in winter, snow-covered slopes.

L.A.'s perpetual sun and cloudless desert skies can perform like

A California sunset bathes San Diego in a warm glow.

San Diego affords cruisers the diversions of both the maritime and the metropolitan.

THE SCENES OF

SAN DIEGO

Despite its reputation for sun and surf, San Diego has a much less ephemeral side, especially when it comes to theater. The Old Globe in Balboa Park, an exact replica of the wooden playhouse that once stood on the banks of the river Thames in London, is the nation's foremost showcase for Shakespeare. Half a dozen productions are presented each year and many are sold out well in advance. The La Jolla Playhouse has a much more contemporary bent, a testing ground for new cutting-edge plays and avant-garde revivals. Many of the productions advance to Broadway fame and fortune, like the hit musical *Rent*, first staged at La Jolla. ⚓

Port appeal rating

Adventure	★★★★
Entertainment	★★★★
Romance	★★★½
Cuisine	★★★½
Shopping	★★★½

THE SIGHTS OF

LOS ANGELES

Don't expect to see movie stars driving along the Sunset Strip or hanging out at Hollywood and Vine. The best places to spot famous faces are trendy restaurants like Spago Beverly Hills and the ever-popular Morton's, or late-night hot spots like the Viper Room and the SkyBar at the Mondrian Hotel. If you haven't got big bucks or cutting-edge frocks, don't fret. You can always eat at a joint owned by some big-name star—Schatzi in Santa Monica (Arnold Schwarzenegger), Thunder Roadhouse, (Dennis Hopper and Peter Fonda) and Eclipse (Whoopi Goldberg and Steve Seagal) in West Hollywood, or 72 Market Street in Venice (Dudley Moore and Liza Minnelli). ⚓

Port appeal rating

Adventure	★★★★
Entertainment	★★★★½
Romance	★★★
Cuisine	★★★★½
Shopping	★★★★

the best stage lighting, illuminating color and shape. The seductive visions sneak upon you, unannounced—a shimmering line of palms, the gleam of a vintage roadster in the next lane over on the freeway, the opulent plumage of a wild peacock preening on the grass. Even first-time visitors may feel a sense of déjà vu, thanks to the city's starring role in countless movies and television shows. The city's cast of characters may look familiar, too: Sharon Stone across the room at a restaurant, Jack Nicholson swinging on a golf course, Jay Leno cruising around Beverly Hills on his flashy Harley.

Yet no matter how many times you've seen it, the City of Angels still holds a few surprises up its stylish sleeve. Sometimes, it's an unexpected whiff of sweet jasmine, or the fresh green aroma of eucalyptus trees. A top-down drive along the coast is a bevy of sensual delights: the intensely warm desert sun on your skin; the cool salt-scented breeze whipping through your hair; the medley of sounds, from crashing waves to the shrill cry of seagulls circling overhead.

The Bradbury building in downtown Los Angeles is a historic landmark.

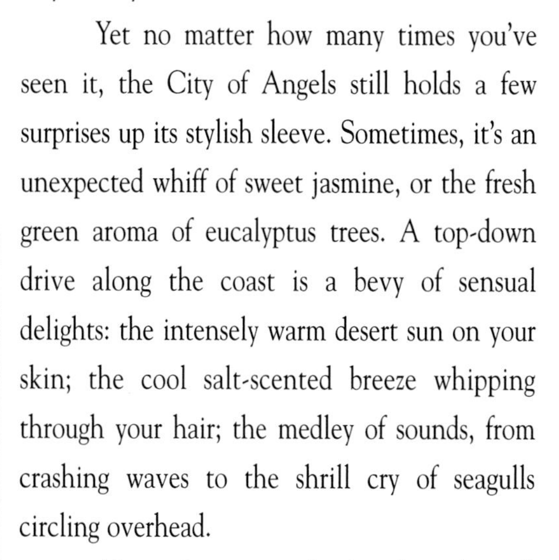

Along the coast, the beach paths roll with a continuous stream of skaters and bikers, walkers and people-watchers. Venice Beach is the epicenter of this beachside pageant, a madcap carnival jammed with fortune-tellers, sidewalk artists, and daring jugglers who toss knives and roaring chain saws into the air.

For a more serene experience, you need look no further than the Palos Verdes Peninsula, jutting out above the cruise ship harbor. Follow the road around water's edge, and you'll see quiet coves scalloped from ocean cliffs, their tide pools teeming with sea life. At its western tip sits the old Point Vicente lighthouse, and an enticing little park set along the cliff, with sparkling views of the Pacific. Whales swim so close to shore that, in winter, dozens are spotted each day, spouting and leaping into the air as if to entertain those gathered on the promontory. At the day's end, it is one of the finest places to watch the skies flame into sunset.

Across the Pacific is Santa Catalina. Catalina's tiny burg is Avalon, tucked cozily into the hills above a horseshoe-shaped harbor. Cars are restricted here; the streets echo with the soft putter of golf carts, a charming contrast to L.A.'s obsession with the automobile. High on the hill sits the Zane Grey adobe, with a commanding Pacific view that inspired its former owner, the prolific Western novelist. Grey is partly responsible for one of Catalina's stranger attractions: a large herd of buffalo that graze on land outside Avalon—leftover props from an old Western movie based on a Grey novel.

Los Angeles freeways are icons and tribulations to Angelenos.

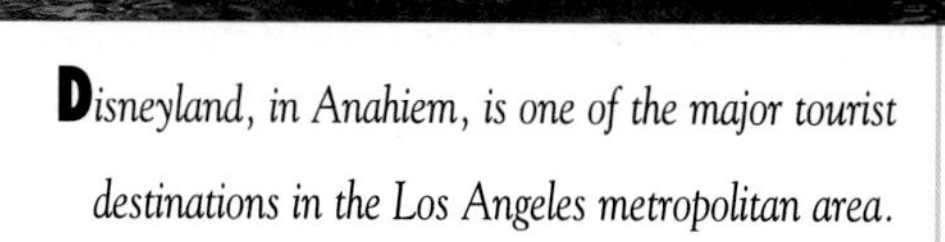

Disneyland, in Anahiem, is one of the major tourist destinations in the Los Angeles metropolitan area.

North of L.A. sits one of the most scenic areas of California—or the world, for that matter: Highway One. From Santa Barbara up to San Simeon, the road veers away from the shore and its string of secluded beaches and sand dunes. But north of San Simeon is the craggy Big Sur Coast, a tangle of thick forests and cool foggy canyons, with jagged peaks that drop straight to the sea. Here, the Santa Lucia Mountains draw tight against the Pacific, so that sometimes you round a cliff, and . . . whew! The glimpse down, down, down to the crashing surf leaves a flutter in your stomach.

North of Big Sur, the mountains recede and the coastline dips in to form Monterey Bay, lined with picturesque towns—quaint tree-lined Carmel By-the-Sea, historic Monterey, and rowdy Santa Cruz, with its beachfront amusement park.

To the north, along the edge of a larger bay, sits one of the world's most beloved cities. San Francisco is a graceful, refined beauty, imbued with enough quirky charm to woo even the most reluctant visitor. The best way to see the city is to hop aboard a cable car, that quaint contraption invented here to help people move around on the city's roller-coaster hills. Stand out on the sideboard in the cool breeze,

The lights of the Bay Bridge shine over San Francisco's waterfront.

THE HEART OF

SAN FRANCISCO

No other American city evokes such romance as San Francisco; visitors lose their hearts there every year. Pretty Victorian houses cling to the sides of steep streets, America's last cable cars rattle up and down the hills, sea lions charm visitors at the seafront, and the colorful Chinese and Italian districts give the city a cosmopolitan flair. Fine cuisine and well-situated restaurants make dining out a romantic adventure. ⚓

Port appeal rating

Adventure	★★★★½
Entertainment	★★★★½
Romance	★★★★½
Cuisine	★★★★★
Shopping	★★★★

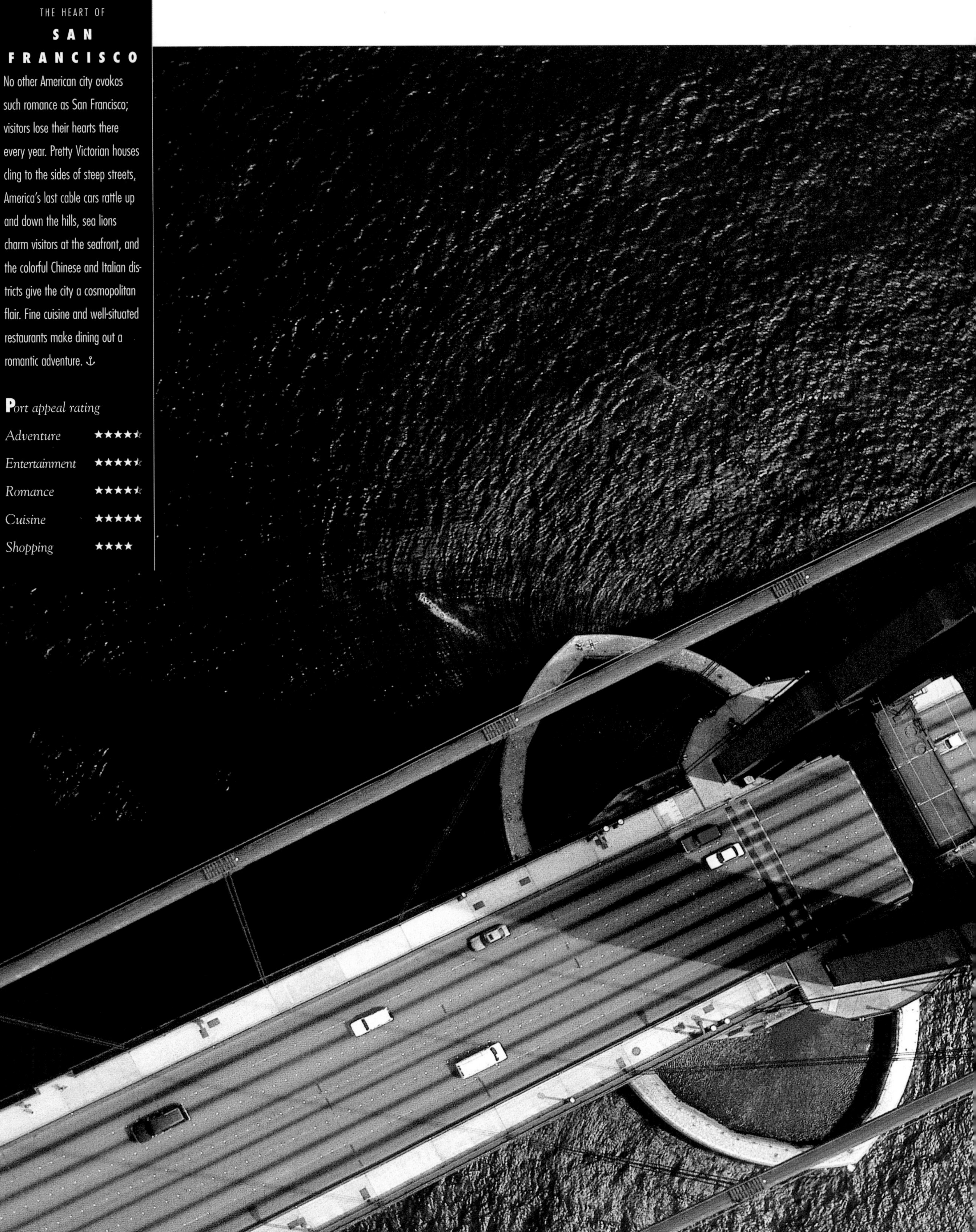

The Golden Gate Bridge stretches a mile across the entrance of San Francisco Bay.

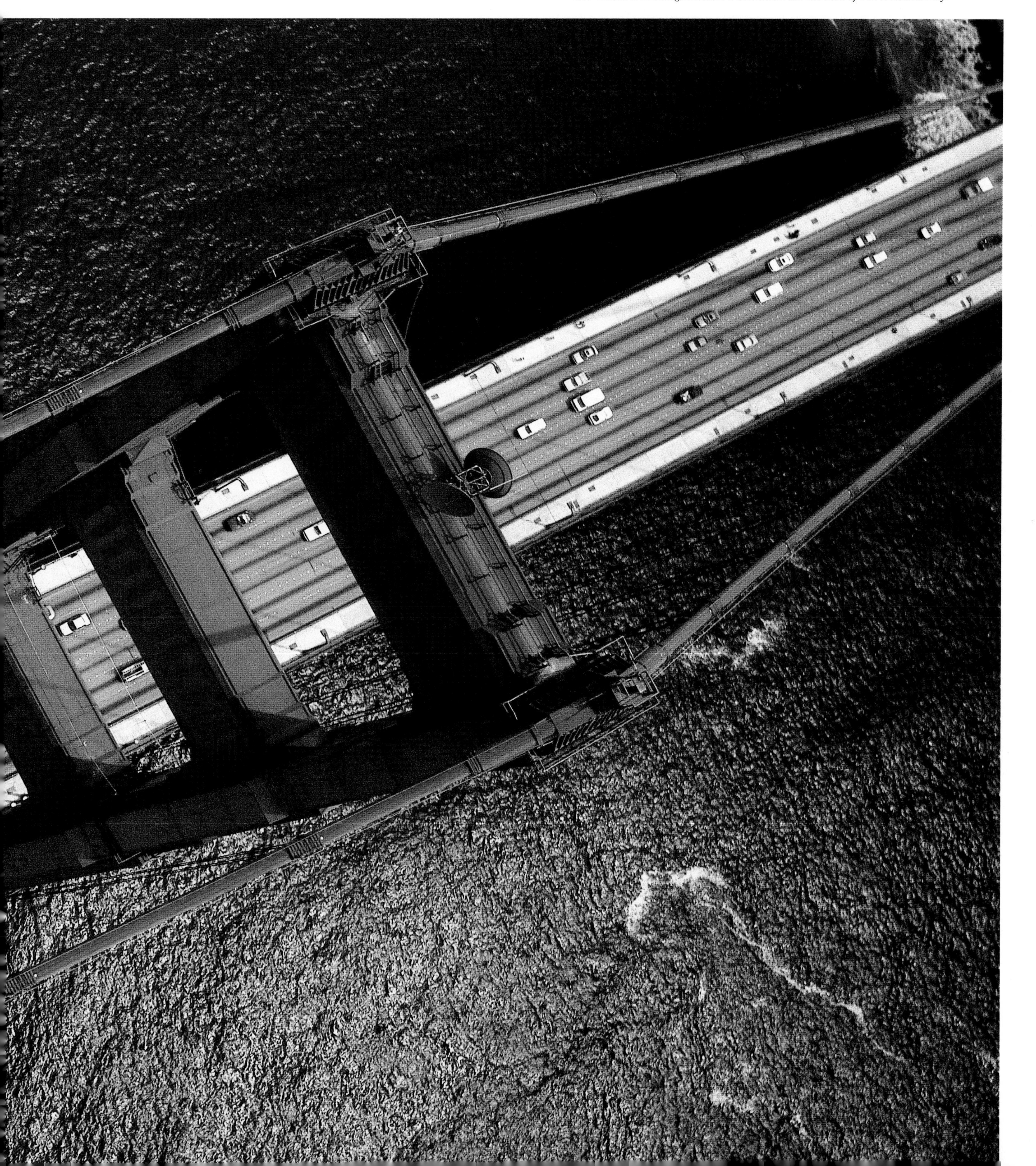

and you can smell the salt water down near the waterfront, the garlic emanating from homey Italian restaurants, the exotic spices of Chinatown, the carefully tended roses near Telegraph Hill.

No matter when you visit, you're likely to hear the foghorn's low boom out in the bay. Fog doesn't creep in on little cat feet here: instead, it rolls in like a thick blanket, sealing off the world softly, dreamily. The swirling mists part, revealing brief glimpses into other worlds: the quiet crowds turning slow-motion tai chi moves in Washington Square; a jogger making his way across Golden Gate Bridge; the sea lions barking hoarsely near Fisherman's Wharf.

On most days, the fog lifts by noon, unveiling sunny skies. Across the Golden Gate Bridge lie two of the area's natural highlights: Muir Woods, a spectacular old-growth redwood grove, and Mount Tamalpais, an otherworldly perch with lofty over-the-cloud views stretching out for miles and miles. On the east side of the bay, in the college town known as the People's Republic of Berkeley, sidewalk

cafés bristle with free thinkers and tea drinkers. More bourgeois pursuits are chased in Sausalito and Tiburon, the pretty bayside towns. Next to the water, artists set up easels to catch the panorama of San Francisco and its bay. By dusk, time seems to stop as people stroll along the shore to gaze at the city, its lights twinkling so romantically across the waters.

Farther north are the sun-soaked valleys of Napa and Sonoma, dotted with hundreds of fine wineries. Wherever there is fine wine, there also tends to be fresh, creative cuisine to accompany it, and these

Below:
W*ashington's Olympic Peninsula flaunts a peerless coastline.*

Overleaf:
L*ions Gate Bridge leaps Vancouver's Burrard Inlet in a single bound.*

THE INDULGENCES OF

SEATTLE

Seattle easily takes the cake as America's trendiest city (cradle of coffeehouses, grunge rock, and e-mail — need we say more?). But the city is also a convenient jumping-off spot for some of the best of Pacific Northwest nature. Mount Rainier hovers above the city, beckoning hikers and climbers with its snowcapped summit. The Cascade Range is less than an hour away, a great evergreen wilderness with finger lakes and granite peaks. Puget Sound harbors killer whales and myriad islands, many of them uninhabited. And there's always the Olympic Peninsula with its rain forest and rugged beaches. ⚓

P*ort appeal rating*

Adventure	★★★★½
Entertainment	★★★★
Romance	★★★½
Cuisine	★★★★
Shopping	★★★½

THE BUSTLE OF

VANCOUVER

Vancouver's Chinatown is the second largest in North America after San Francisco. The heart of the district is Pender Street, lined with Asian restaurants, grocery stores, and souvenir shops. The Chinese Cultural Center along this stretch offers an ongoing slate of classes, lectures, cultural performances, and art shows with Asian themes. Directly behind the center is the Sun Yat-sen Chinese Garden, designed by artisans and landscape architects from Suzhou and the first of its type outside of China. ⚓

P*ort appeal rating*

Adventure	★★★★½
Entertainment	★★★★
Romance	★★★★
Cuisine	★★★★
Shopping	★★★½

Mount Rainier is a dramatic backdrop to downtown Seattle

valleys are no exception. One of the greatest pleasures here is to gather a picnic—fresh cheeses, breads, fruit—and slip away to a shady spot, preferably one overlooking rows of grapevines and the golden valley beyond.

Above San Francisco, the coastline grows rockier and wilder, more isolated and lonely. Redwood forests tower for miles on end, then fade away into the marshes and sand dunes of Oregon.

To the northeast, far from the warm Pacific, sits Seattle, poised prettily between the Olympic Mountains and the Cascades. The mountains do more than provide scenic backdrops and outdoor adventures: the Cascades block wintry blasts from the east, while the Olympics shield the city from chilling rains, which instead drench the Olympic Peninsula rain forest. Seattle gets its nickname, the Emerald City, from its leafy ceiling of trees overhead, and the grassy parks dotted throughout the metro area. But there's also that panorama of downtown's futuristic skyscrapers, so striking from hilltop parks, that gives it a resemblance to the fantasy Emerald City of Oz.

One need not board a balloon to escape from this wonderland. Seattle sits astride an inland waterway so vast that its first European visitor mistook it for an ocean. Puget Sound's waterways stretch from below Seattle all the way to Canada. Hop in a boat, or board a ferry, and you could cruise for days, stopping at idyllic islands that dot the Sound.

Seattle's calm waters are great for kayaking, and its miles of bicycle paths allow one to get a closer look at neighborhoods nestled along the shore. Downtown's Pike Place Market is not to be missed: this open-air collection of produce stalls, fish markets, and handicraft stands is also a great place to grab a steamy cup of joe and sit overlooking Elliott Bay.

Totem poles grace Brockton Point in Vancouver's Stanley Park.

Vancouver and Seattle might have been twin cities, yet they took different paths to growth. As Seattle expanded out for miles around Puget Sound, Vancouver centered its future on its slim finger of land. Today, Vancouver is a hip, sophisticated city of high-rises, its urban density second only to Manhattan in all of North America. Somehow, it has retained its natural beauty and its friendly, small-town coziness. The streets are lined with trees and flowers; the surrounding waters remain fresh and clean. The central green space, Stanley Park, is a wildly beautiful oasis with thick forests, stunning views, and pathways always humming with joggers and cyclists. In a clearing, a gathering of authentic totem poles stands solemnly; in a glen, a rose garden blooms profusely.

Just across the bay is Grouse Mountain, overlooking the bay and the city. Though Vancouver's air rarely drops below freezing, Grouse rises high enough to be covered with snow for most of the winter, a delight for skiers who can go from city to slopes in minutes. Despite its proximity to the city, this is not some urban-squeezed refuge: bears roam the woods here, and the trails roll for miles into the forest. During summer, hikers scramble up the steep hill to the summit. Less active folks simply board the tram for a nearly vertical ride to the top. Descending can also be an adventure: daring parasailers glide down to the base, skimming treetops with their butterfly-colored contraptions.

Exotic plants survive the Canadian winter beneath the dome of Bloedel Conservatory in Vancouver.

Vancouver is the starting point for most cruises into Alaska's Inside Passage, a narrow inland waterway blessed with calm waters and pristine scenery. As the boat travels north, the mountains rise higher and higher, their dramatic peaks layered majestically on the horizon. Wildlife—bears, otters, whales, and porpoises—is abundant along these channels, as is the gentle thunder of waterfalls plunging through untouched forests.

Right: A Tudor-style village adds to Victoria's delightful British flavor—a part of the attraction for its many appreciative visitors.

Overleaf: The splendid colors of the Burchart Botanical Gardens on Vancouver Island are just a short drive from the city of Victoria.

Alaskan ports of call are still frontier towns, rugged and rowdy and filled with colorful history. Many, including the state capital of Juneau, are linked to the outside world only by air or the Inside Passage. In the 1800s, when John Muir sailed here, he bemoaned the crowds who rushed off the boats to buy souvenirs that barely represented the Alaska he knew and loved. It is no different today. Yet today's passengers can also, in mere minutes, experience the wild Alaska it took Muir days to reach from town.

Helicopters whisk adventurers to the nearby glaciers, where they can try dogsledding or trekking along the jagged blue ice peaks—without the misadventures of Muir, who sometimes fell into the crevasses. Rafters travel paved roads to the raging rivers, but they continue to see plenty of bald eagles and bears roaming the banks.

Glacier Bay is one of the highlights experienced on an Alaska cruise.

Still, you don't have to go far from port to experience Alaska's wonders. Every town has at least a couple of hiking trails leading into the deep woods. In Ketchikan, the town's main creek is choked with silvery salmon attempting the treacherous upstream journey. Juneau's famous resident glacier, Mendenhall, glides and groans along the edge of the city. In Anchorage, if you take a bicycle ride, you may be warned to watch out for gangly moose. Bears are everywhere, though they are shy and difficult to see—unless you live near the woods in town.

The whole town of Skagway is one big historic landmark that hums with activity when cruise ships dock. It is also home to one of the world's engineering marvels: a narrow-gauge railroad that snakes up an impossibly steep trail used by prospectors during the Klondike gold rush. From town, the train rolls past a glacier-fed river, then chugs upward along a forested valley, through dark tunnels and past waterfalls.

The Tlingit people are responsible for carving most of Alaska's totem poles.

The Gold Rush town of Skagway is the final stop on Alaska's Inside Passage.

One of Alaska's favorite stops is not even a port, but a journey into another world, where glaciers continue to sculpt the landscape, much as they did eons ago farther south along the West Coast. Ships can draw close to the face of these rivers of deep blue ice, close enough to see seals sprawled out on the bergs and birds that fly en masse at each crash of a berg. The mists from the bay rise eerily, and the depth of quiet is broken only by the low moan and crackle of the ice before it plunges into the bay.

At the end of any day, any-where along the West Coast, your attention will naturally turn toward the west, to the setting sun. No matter where you are, from San Diego to Anchorage, you won't be far from a proper theater—a rocky promontory high above a sweep of peaceful coves, a windswept beach lapped by flaming waves, a mountain studded with redwoods rising above the clouds, a quiet blue bay opening to the Pacific. On most days, the skies alight in a spectacular clash—the gentle blue bowing to the majestic scarlets and tangerines, violets and golds—that flares and blares with an intensity that can't be ignored. All too soon it fades away, as the amethyst skies of the evening emerge, darkening slowly to cobalt against a silvery moon. ⚓

Denali (Mount McKinley) towers 20,320 feet above central Alaska.

BY ROBERT W. BONE

O·C·E·A·N·I·A

A WORLD OF WATER

MOST AMERICANS COME INTO Oceania at Honolulu, a good choice, but a pity the way they do it. In the drone of a few jet hours, they hardly have time to turn their watches back a few time zones, let alone turn their way of thinking from the hustle and bustle of the mainland to the more languorous moods of the Pacific. Luckier are those who arrive in the land of Aloha the way it was done before World War II. After four or five days relaxing on the open sea, passengers gather on the upper decks of a grand, white-bottomed vessel. There they vie good-naturedly to see who can be the first to catch sight of the distinctive promontory named Diamond Head.

Diamond Head is the southern rim of an extinct volcanic crater. Before modern communications were available, fires lit on its summit announced to Honolulu that an incoming ship had been sighted. For thousands of years Diamond Head has stood guard over a decorous curve of golden sand known as Waikiki Beach. And for more than a century, it has also served as a dramatic symbol for the Hawaiian Islands. Indeed, variations of its familiar volcanic form are used by artists in combination with sand, surf, a carefully placed palm tree, and perhaps a young woman, wearing a grass skirt and strumming a ukulele, to create a scene evoking not just Hawaii, but any typical South Seas paradise.

Islands, and the volcanoes or coral reefs that created them, are the essence of Oceania. There are at least twenty-five thousand of these green and golden Elysiums, more than half mixing balmy trade winds and salty air with the fragrance of coconuts and flowers like jasmine and the ubiquitous frangipani. The very existence of a Pacific Ocean was not known to Europeans for centuries. When Ferdinand Magellan, James Cook, and others returned home from expeditions with tales of wonder, they were met by some with gales of derisive laughter. "Come, come now! A creature the size of a greyhound and who leaps like a grasshopper? What do you take me for!" Then when they were told that some kangaroos even come equipped with a pocket . . . well, the reaction in the corner pub was not always sympathetic. Nevertheless, the imaginations of new generations of

BRISBANE
HOBART
HONOLULU
MELBOURNE
PAPEETE
PERTH
SYDNEY

The cultural influences of the Aborigines enrich the travel experience through the Australian Outback.

romantics began to run wild. Some decided to face the unknown dangers so they could see for themselves. These adventurers found even more strange plants and animals, along with new and interesting peoples to trade with, to fight, and to love. Explorers were followed by novelists, poets, artists—even actors. Pacific islands became the world of Herman Melville, W. Somerset Maugham, Robert Louis Stevenson, Jack London, James Michener, Paul Gauguin, Dorothy Lamour, and perhaps even Willy Gilligan on his "three-hour tour." Some came to adore their newfound Pacific paradise with religious fervor. "This serene Pacific, once beheld, must ever after be the sea of his adoption," said Melville's Ishmael in *Moby Dick*.

Among the cultures of Oceania, Polynesia is the best known. It is often described as a triangle, with Hawaii at the pinnacle; Easter Island and its mysterious stone faces to the east; and Aotearoa, what the Maoris called New Zealand, at the bottom. Also enclosed within the triangle are the Polynesian peoples of Samoa, Tonga, the Marquesas, and, of course the islands of Tahiti, birthplace of Polynesia.

Anthropologists also have defined Melanesia as including the islands of Fiji, New Caledonia, and several others. And then there's Micronesia, scattered groups of civilizations farther west, many of them well known as battlegrounds in World War II. Today these are more famous for their multicolored fish and exceptionally clean and clear water, attracting many of the world's scuba diving fanatics. Even Bikini, which symbolized the atomic age and gave its name to a bathing suit, has been cleaned to the extent that recreational divers descend safely to view some of the ships that were sunk there during hydrogen bomb tests in the 1950s.

Since 1925, the Aloha Tower in downtown Honolulu has greeted cruise ships to what in 1959 became the capital of the American state of Hawaii. The ten-story structure, once the tallest building in the Hawaiian Islands, bears the Hawaiian word that means *welcome*, *good-bye*, and *good wishes* all at the same time. Recently, an attractive shopping, museum, and entertainment complex was created around the Aloha Tower, and those whose ships dock at Piers 9 or 10 will find themselves pleasantly located. Waikiki Beach is a short taxi ride away. Those staying overnight have their choice of luxury hotels and plenty of opportunity to experience a luau or other enchanting Polynesian entertainment.

Some may even learn to do the hula or play the ukulele. Friendly locals may even teach a malihini (newcomer) a Hawaiian word or two. Although few really speak the language, some ancient words regularly spice up the local lingo, like *mahalo* for *thank you*. The sites of two famous battles can be visited either by tour bus or by rental car. The first, of course, is the USS *Arizona* Memorial at Pearl Harbor, the scene of the 1941 attack that plunged the United States into World War II. Near Honolulu, the Nuuanu Pali Lookout provides a spectacular view over ocean and valley. This was also the scene of a decisive engagement in 1795, when the conquering King Kamehameha forced most of the defending army over the steep cliff, thus consolidating his claim to a wider realm.

Hawaii's native heritage includes woodcarving (left) and traditional dance (above).

Pronounced *pango-pango*, the city of Pago Pago is the capital of the U.S. territory of American Samoa. Two large islands farther west compose the independent nation of Samoa.

There is considerable traffic as well as a fair amount of rivalry between the two Samoas, and many families are separated by them. Most residents of American Samoa are bilingual, speaking English along with Samoan. As a result of intense missionary activity in the nineteenth century, nearly all Samoans are devout Christians. A Sunday visit to a Samoan church, with its choir singing in

Sunrise casts an orange glow across the Hawaiian coast.

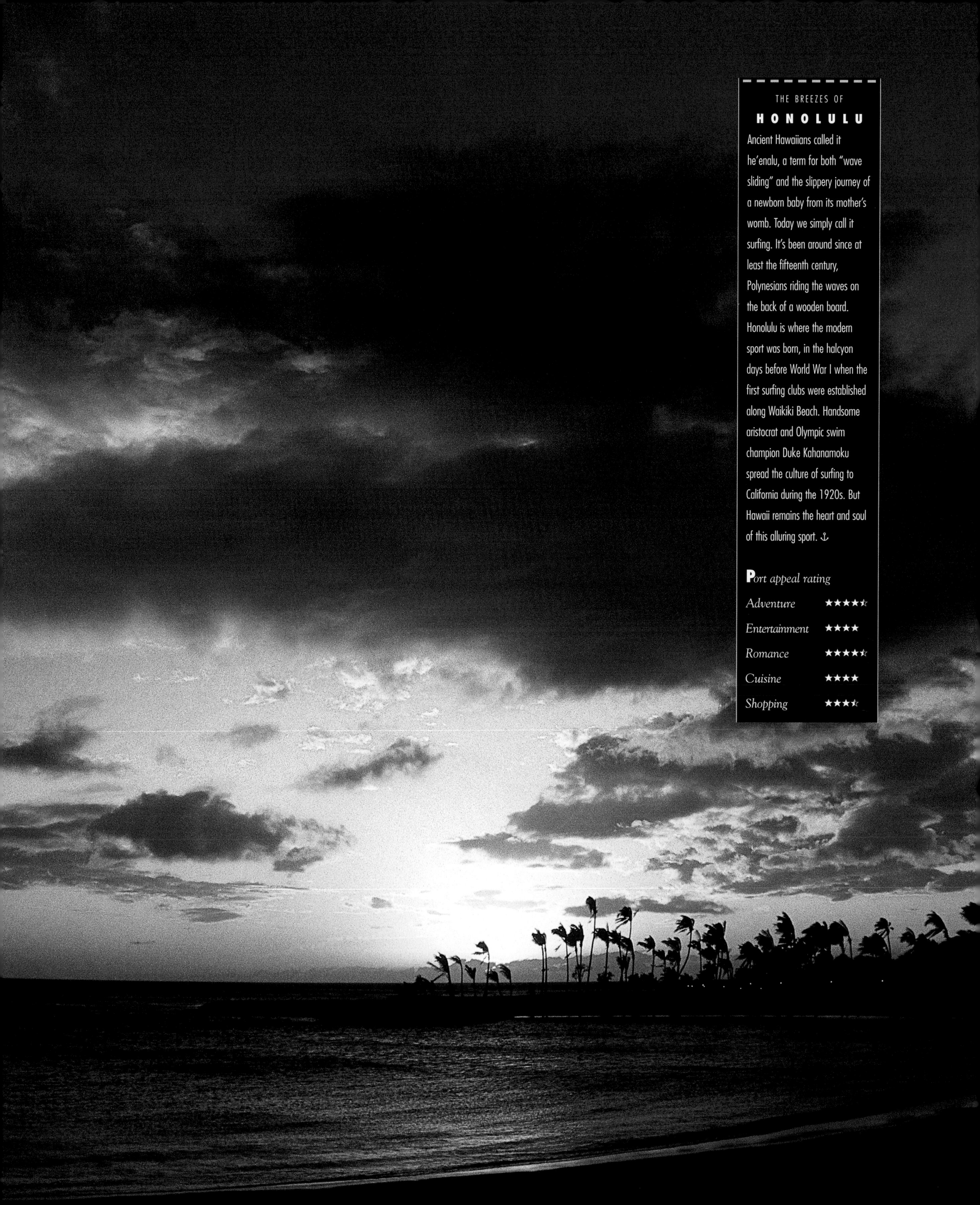

THE BREEZES OF

HONOLULU

Ancient Hawaiians called it he'enalu, a term for both "wave sliding" and the slippery journey of a newborn baby from its mother's womb. Today we simply call it surfing. It's been around since at least the fifteenth century, Polynesians riding the waves on the back of a wooden board. Honolulu is where the modern sport was born, in the halcyon days before World War I when the first surfing clubs were established along Waikiki Beach. Handsome aristocrat and Olympic swim champion Duke Kahanamoku spread the culture of surfing to California during the 1920s. But Hawaii remains the heart and soul of this alluring sport. ⚓

Port appeal rating

Adventure	★★★★½
Entertainment	★★★★
Romance	★★★★½
Cuisine	★★★★
Shopping	★★★½

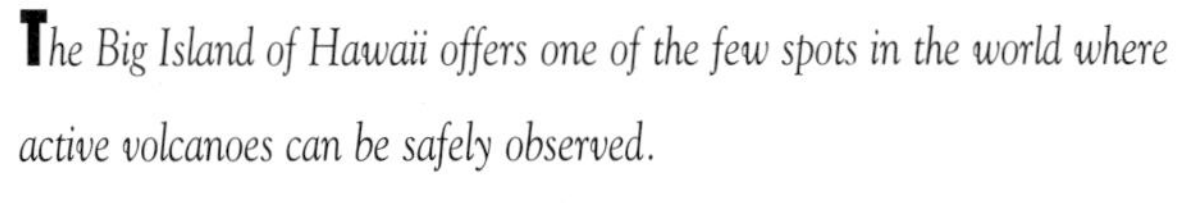

The Big Island of Hawaii offers one of the few spots in the world where active volcanoes can be safely observed.

Right: The sunshine of the summer months Down Under—December, January, and February—warms up visitors from the wintering north.

Native island ceremony and pageantry are among the thrilling spectacles awaiting visitors to the Hawaiian islands.

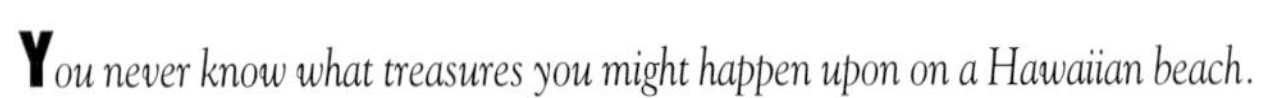

You never know what treasures you might happen upon on a Hawaiian beach.

Man-made and verdant natural beauty harmonize throughout the paradise of the Hawaiian islands.

THE WARMTH OF

PERTH

Perched on the banks of the Swan River, Perth bills itself as the sunniest capital city in Australia. The warm, dry, Mediterranean climate is ideal for growing grapes, one of the major crops of the Swan River Valley east of the city center. The first wineries were established in the 1840s at around the same time as those of the Barossa Valley of South Australia. Most Perth growers are more than glad to show you around the property and render a few sips of their latest vintage. A half-day cruise in the Swan River is a good way to tour the wine country. ⚓

Port appeal rating

Adventure	★★★½
Entertainment	★★★
Romance	★★★½
Cuisine	★★★
Shopping	★★★

THE COLORS OF

PAPEETE

Paul Gauguin, a successful French stockbroker, turned his back on the financial world in 1891 and set sail for Papeete, where he hoped to find the peace and happiness that had eluded him in Paris. With a young Polynesian mistress named Tehura, he retreated to a small wooden house on the south coast of Tahiti where he spent the rest of his days living like a native and rendering his tropical Eden on canvas. Family and friends back in Paris thought he was loopy. Yet Gauguin was perfectly content, one of the first outsiders to discover what the Tahitians themselves had long known—gratification can be as simple as an empty beach, a cool breeze, and a beautiful woman. ⚓

Port appeal rating

Adventure	★★★★
Entertainment	★★★
Romance	★★★★
Cuisine	★★★
Shopping	★★½

Cook's Bay provides shelter for yachts and cruise ships visiting Moorea island near Tahiti.

close, multipart harmony is truly an unforgettable experience.

A *new day dawns on Tahiti.*

Pago Pago's most well-known missionary is the repressed cleric who converted—then seduced—Miss Sadie Thompson. Travelers are pointed to more than one building said to be the home of the unfortunate lady described in Maugham's moody story, and played by Rita Hayworth in the movie *Miss Sadie Thompson*. The traditional sightseeing route includes a morning trip into the rugged green mountains above the capital, affording panoramic views of the harbor and the ships anchored there. Some follow this with lunch at Sadie's Restaurant, an atmospheric address that may or may not have been the home of Miss Thompson herself.

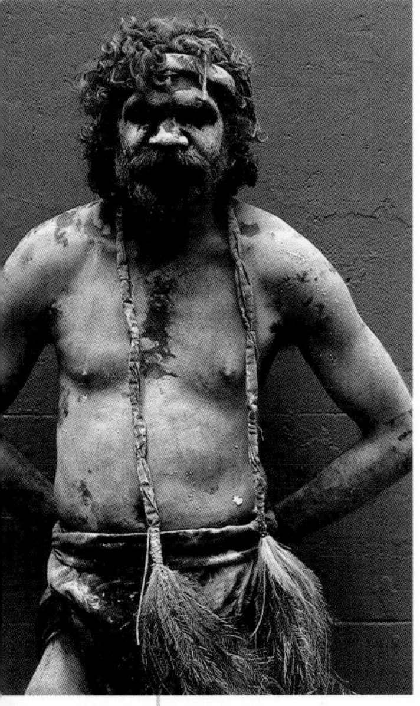

M*eeting aborigines—linked by religion to both land and nature—is an integral part of the travel experience of Australia.*

On the island of Tahiti, the capital of French Polynesia, Papeete, still evokes some of the same mysterious magnetism that has drawn artists like Paul Gauguin and other seekers of an unhurried pace of life to its shores for nearly two centuries. The music, song, drums, dances, and friendly people that mesmerized the mutineers of the HMS *Bounty* can still be seen and heard in Tahiti. Add to that the gastronomic wonders that any French culture seems to produce, and Papeete adds up to a wonderful place for food and entertainment. Cruise ships often tie up right in downtown Papeete, making it especially convenient for shopping and sightseeing. If the city seems too frenetic, travelers can take a short hop across the channel to the beautiful and slow-paced neighbor island of Moorea, allegedly the inspiration for Michener's Bali Ha'i.

Tahiti has its share of cuisine to sample and souvenirs to purchase. For an appetizer, visitors should try poisson cru, the local fish (tuna or bonito) marinated in lime juice, a delicious specialty of Tahiti. Its tangy flavor seems to linger, influencing the main course that comes later. Women will surely want to pick up an alluring pareu or two. The price may also include a lesson in exactly how to wrap it so it stays on during a hip-swinging hula. The best ones are colorfast; some cheaper ones might not be. Tahiti is also the traditional place to shop for genuine black pearls, which, of course, are not completely black. (Chances are any totally black ones are Japanese and dyed that way, and are in danger of fading.) The best are often gray, greenish gray, or bluish gray.

Hawaii, Samoa, and Tahiti may be easily recognized as traditional capitals of Polynesia. Less obvious, perhaps, but no less a link in the same culture is New Zealand—called Aotearoa, the Long White Cloud in the Maori language.

At first glance, Auckland, the country's largest city, is unabashedly British. Women dress in some of the latest London fashions; fish and chips shops abound; statues celebrate military heroes from two world wars. New Zealanders, who call themselves kiwis, speak English with a British beat. But New Zealand is actually a blend between two cultures that were often at war with each other one hundred years ago. Today Maori and Pakehas (non-Maori) share a fiercely egalitarian government. Still, many young Maori feel discriminated against in subtle ways. In concert with indigenous peoples in other areas of the world, they are asserting their own cultural roots more than ever.

Travelers who sail into Auckland on a weekend share an attractive harbor with hundreds of sailboats, all piloted by locals enjoying a respite on the water; indeed, Auckland likes to call itself the City of Sails. For a demonstration of kiwi conviviality, a foreign traveler should step into a local pub and order perhaps a pint of Lion Brown. A non-New Zealand accent will quickly spark a friendly conversation with a fellow patron. Auckland is also an excellent place to shop for sheepskin rugs, but they should be bought only in a shop that specializes in the product. An honest proprietor will cheerfully explain the differences in quality.

Many visitors opt for a day trip to Rotorua, considered the center of the Maori culture. A few Maori still cook in pots that boil in the hot soil of Rotorua. Scattered among the sulfurous steam and dramatic thermal activity there, the indigenous people of Aotearoa stand ready to introduce newcomers

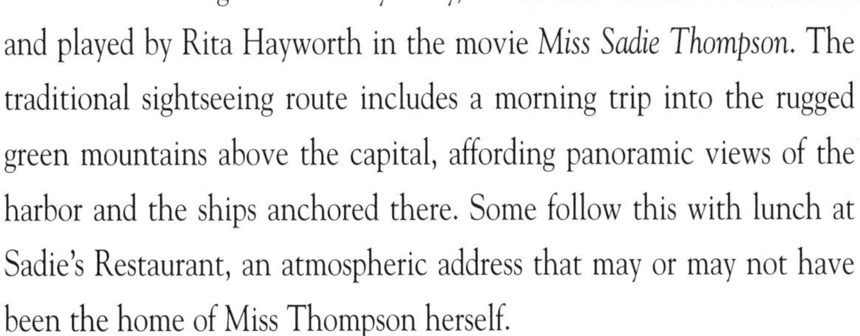

C*arved Polynesian idols reflect the rich spiritual and artistic heritage that so fascinates visitors.*

to their arts, music, dance, and other samples of a way of life that has all but disappeared in modern-day New Zealand. A hangi, or feast, includes music, dances, and other demonstrations of rhythmic ability. Today some—but not all—of the tattoos the Maori wear on chins and arms will wash off when the show is over. The last song, which was once popular in the United States as "Now is the Hour," is actually a Maori hymn that often brings tears to the eyes of sentimental Maori and Pakehas alike.

The principal city on the South Island of New Zealand, Christchurch, is not really a port at all. It is separated by a ridge of atmospheric hills from Lyttleton, which lies on the water. Despite its urban environment, Christchurch enjoys a special, peaceful mood, largely created by the meandering River Avon. It winds here and there through the city, always bordered by grassy banks that serve as pleasant picnic grounds for ducks and office workers alike. Visitors can be taken punting on the Avon. Many just wander for much of its length on foot or by bicycle.

Nearby is Cathedral Square, an open area forbidden to vehicular traffic, and dominated by a century-old Gothic cathedral. It's a university town, at times filled with legions of cavorting students. New Zealanders like to call Christchurch the Garden City, and a veritable festival of colorful blossoms provides the evidence. Local floral contests frequently make homes, schools, and even factories bloom with pride throughout the city.

Australia weighs in as the Big Daddy of the South Pacific. As

Brisbane offers easy access to both the Gold and Sunshine coasts of Queensland.

THE TREATS OF

BRISBANE

Queensland's capital is the gateway to some of Australia's most spectacular outdoor attractions. The endless beaches of the Gold coast and Sunshine coast flank the city on either side. Lone Pine Koala in suburban Fig Tree Pocket is an outstanding place to have close encounters of the marsupial kind—kangaroos, wallabies, platypuses, wombats, and over a hundred cuddly koalas. Inland are great rain forest preserves like Tamborine Mountain and Lamington National Park. And then there's the Great Barrier Reef, the world's single greatest underwater attraction. The southern edge of the reef lies around Rockhampton, about an hour's flight north of Brisbane. ⚓

Port appeal rating

Adventure ★★★★
Entertainment ★★★
Romance ★★★
Cuisine ★★★
Shopping ★★½

Ayers Rock is a sacred shrine to aborigines living in Australia's Red Centre.

isolated as it may seem on the globe, its size is comparable to the forty-eight contiguous states of the United States. Its great cities, however, occupy mostly the rim of the island continent, since much of the country's interior is considered a vast desert, generally known as the Outback. Not that there isn't a certain charm about what Aussies also call the Red Centre. That's the location of Ayers Rock and the storied Alice Springs.

Sydney is where Australia began with a small colony of prisoners shipped out of Mother England in 1788. The country's convict beginnings are said to be responsible for the breezy, generally anti-authoritarian attitudes of the population still today. After fighting side by side in two world wars, Aussies seem to have a special affinity for "Yanks" and are usually eager to show genuine Down Under fellowship at every opportunity. An equally friendly American traveler is more likely to be invited home for dinner in Australia than almost anywhere else on earth.

The harbor at Australia's largest city boasts a long shoreline of coves and bays that effectively display the charms of the neighborhoods that occupy its banks. Today Sydney is becoming almost as well known for its architecture as for its harbor. Sydney's grand Opera House, with its majestic sail-like roof, occupies a dramatic site across Circular Quay from the passenger terminal. Ship passengers like to disembark and then go on a pub-and-shopping crawl in the Rocks, the atmospheric, old-world neighborhood nearby. A cruise on the harbor, by ferry or hydrofoil, is also popular.

In the city center, near Sydney Tower, sometimes named the Golden Bucket by iconoclastic Sydney-siders, modern-day explorers take the monorail to another renovated fun neighborhood called Darling Harbour.

Sydney's traditional rival in sports and in nearly everything else is the wonderfully sophisticated city of Melbourne, about five-hundred miles to the south. Many in Sydney would admit that Melbourne holds the lead in fashion, theater, and restaurants. It is also an excellent place to shop for opals, although stores nearly anywhere in Australia offer these stones for much lower prices than in the United States. Travelers should ask for lots of advice on opal buying and probably not take the first ones offered.

Melbourne owes its beginnings to the gold rushes of the nineteenth century, and the ornate buildings that line leafy Collins Street give elegant testimony to that affluent period in history. It's also one of the few cities in the world with a large, efficient network of tram cars, including some ancient models that clatter and clang their way through busy streets. One of these is a rolling restaurant, which somehow manages to serve gourmet meals while crossing from one end of Melbourne to the other without spilling the wine. Visitors to the city are drawn to the Victorian Arts Center, a dramatic clutch of modern structures alongside the Yarra River. ("Victorian" refers to Melbourne's role as the capital of the State of Victoria, not to a style of art.) Although performances of all types are given in the Arts Center, visitors are also charmed by some of the more traditional stages in town. One of these is the wonderfully refurbished, century-old Princess Theater on Spring Street, a perfect stage for period productions, where *Phantom of the Opera* was a sellout for two years. Many New Zealanders were even willing to pay plane fare from Auckland to Melbourne for the privilege of scaring themselves silly when it seemed the theater's own chandelier was crashing down upon them, only avoiding their heads at the last second to land on the stage instead.

Australia's Outback presents travelers with unique road hazards.

Also an east-coast city, but much farther north, Brisbane differs from both Sydney and Melbourne largely because of its especially mild climate. It's the capital of the tropical state of Queensland, a land of sugarcane, pineapples, bananas, mangoes, and some less-familiar fruits and vegetables. If there seems to be something of an American influence in Brisbane, it may date from World War II when thousands of American GIs were stationed in the city, and many ended up marrying Brisbane girls. General Douglas MacArthur also made his headquarters in the city until the Philippines were liberated.

Since it opened in 1973, the Sydney Opera House has helped to attract both art and music lovers to a country which values both its native and modern arts.

THE TASTE OF

SYDNEY

Sweeping tides of immigration during the late twentieth century—Greeks, Italians, Yugoslavians, Pacific Islanders, Indonesians, Arabs, and Asians—have transformed Sydney from an isolated Anglo-Irish enclave with an insular point of view into a cosmopolitan melting pot where Ramadan and Chinese New Year are as much a part of local life as Easter and Anzac Day. In fact, nearly 40 percent of Sydney's citizens were born overseas. This globalism is especially evident in the city's restaurant scene, which includes everything from kebabs, satay, and sushi to raznjici, moussaka, and chicken vindaloo. You could literally eat your way across Sydney. ⚓

Port appeal rating

Adventure	★★★★½
Entertainment	★★★★½
Romance	★★★★
Cuisine	★★★★
Shopping	★★★½

THE BEAUTY OF

HOBART

It's often said that Tasmania is the most British part of Australia, and in many respects Hobart bears the mien of an English coastal town. Built around a bustling little harbor that was once a major whaling port, the downtown area is adorned with squat Georgian buildings rather than the sleek high-rises that dominate most Aussie cities. A local landmark called the Theatre Royal has hosted the likes of Noel Coward and Sir Laurence Olivier. Fish and chips—with a pint of ale—is a typical waterfront lunch. Even the weather seems decidedly English: damp, cold winters remind you that Hobart is the southernmost city in Australia. ⚓

Port appeal rating

Adventure	★★★★
Entertainment	★★★
Romance	★★★½
Cuisine	★★★
Shopping	★★★

Hobart, Tazmania, glistens peacefully in the afternoon sun.

Koalas rarely descend from their eucalyptus perches.

Brisbane has a fairly large population of urban Aborigines, as compared to Sydney or Melbourne, where they are relatively unnoticed. First-timers to Brisbane should not miss the Lone Pine Koala Sanctuary, a few miles upriver from the city center. Wombats, wallabies, dingoes, and other local critters are also in residence there.

Perth, Australia's jewel in the far west, is actually a short way up the Swan River from Fremantle, which serves as the port for Perth. Back in 1987, Fremantle drew world attention as the venue for the America's Cup yacht races, and it is an attractive town itself. But Australians, who argue a lot about Sydney, Melbourne, and other Down Under metropolises, seem to agree that Perth is almost everyone's favorite city. Known as the sunniest capital in Australia, it enjoys almost perfect weather. Most days, a little after noon, just when the day seems to be building up to be a scorcher, a fresh sea breeze known as the Fremantle Doctor gently begins to soothe the city.

Cruise ships are often welcomed in several other cities on the coast of Australia, notably in Adelaide, Cairns, and Darwin. Adelaide, in the center of Australia's most well-known wine country, is the nation's best-planned city. Its streets and gardens were laid out by the same architect who later turned his talents to Christchurch, New Zealand. Travelers who want to chuck a kangaroo under the chin should find their way to the Cleland Wildlife Park in the nearby Adelaide Hills.

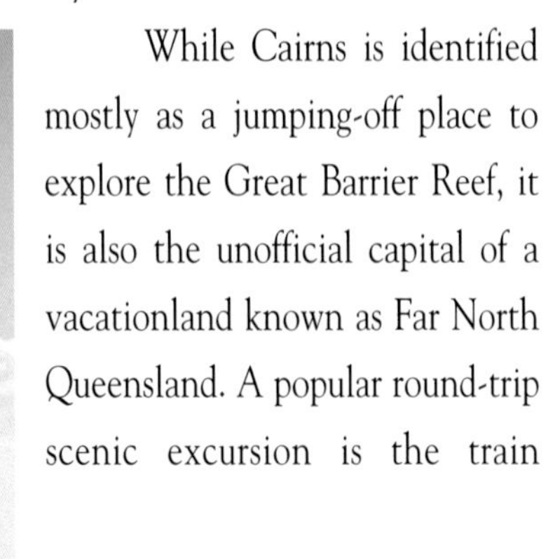

While Cairns is identified mostly as a jumping-off place to explore the Great Barrier Reef, it is also the unofficial capital of a vacationland known as Far North Queensland. A popular round-trip scenic excursion is the train between Cairns and the village of Kuranda high up on the edge of the Atherton Tableland.

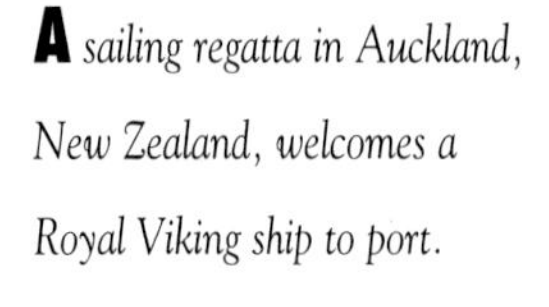

A sailing regatta in Auckland, New Zealand, welcomes a Royal Viking ship to port.

Micronesia boasts the world's largest concentration of coral atolls.

Darwin, though often hot and humid, in some ways is the most different of all the country's ports. The only Australian city to be bombed during World War II, it still shows off a few scars to prove it. But Darwin's main claim to fame is as a gateway to the Australian Outback. Here travelers may be fortunate enough to see a traditional Aborigine corroboree, or celebration, and perhaps to taste a water buffalo steak. Darwin also serves as a launching pad to such animal-rich experiences as Kuranda National Park. The nature cruise on the park's Yellow River, home of crocodiles, cockatoos, and other creatures, is one of the most exciting experiences in the country.

In the same way that explorers began colonizing Oceania a century or two ago, modern-day cruisers have turned their inquisitive noses to vast waters unmentioned in the Bible, unimagined by Homer, and unruled by Poseidon. Those who follow the world of cruising have said that the Mediterranean is the sea of the past, the Caribbean is the sea of the present, and the Pacific is the sea of the future. ⚓

Lawn bowling is popular fun and part of the local color in Sydney.

Despite its modern skyline, Melbourne is considered the most traditionally "British" of Australia's big cities.

THE EXCITEMENT OF

MELBOURNE

San Francisco and Sydney are often mentioned in the same breath, but California's city by the bay actually has more in common with Melbourne. Victoria's state capital was little more than a small fishing port until gold was discovered in the 1850s. Within a decade, Melbourne had boomed into a big city that supplied the gold miners and absorbed much of their profits. Millionaire prospectors embellished the city with extravagant Victorian mansions and local businessmen developed a thriving downtown area that's still called the Golden Mile. The precious metal petered out long ago, but you can visit relics of the gold rush at Ballarat and other mine towns west of Melbourne. ⚓

Port appeal rating

Adventure ★★★★
Entertainment ★★★½
Romance ★★★
Cuisine ★★★½
Shopping ★★★

BY GARRY MARCHANT

F·A·R E·A·S·T

AN ORIENTAL ODYSSEY

EUROPEAN EXPLORERS AND MERCHANTS who first sailed into Asian waters were dazzled by what they saw. In these faraway seas, they encountered ancient civilizations unlike anything in the West, with fabulous cities and fantastic ruins. They marveled at colorful religions with ornate temples and shrines, sultans' palaces, domed mosques rising from the steaming jungle, and people living in houses on stilts over the water. Ladies wore elaborate silk kimonos in Japan and gracious sarongs with gold thread woven through in Southeast Asia, while in the jungles of Borneo, Malaysia, they were elaborately tattooed from neck to knee. In the countryside, women in umbrellalike cone hats worked the rice paddies alongside giant, gentle, mud-slathered water buffalo with great, spreading horns. In temple halls across Asia, rows of monks in saffron robes chanted and prayed.

The Europeans' ships anchored in scenic bays with outrigger canoes pulled up on sandy beaches, and in harbors crowded with sampans, junks, and other ancient craft. These early visitors also found fantastic natural wonders in the vast jungles, strange birds such as great hornbills with colorful "helmets," and animals such as rhinoceroses, elephants, proboscis monkeys with pendulous noses, comical orangutans, and giant monitor lizards. And merchants were attracted by "green gold"—spices with fabulous new flavors—that made them fabulously rich. The mystical East was the splash of water in a shrine fountain, the tinkle of temple bells, the rich scent of sandalwood incense. This great land of mystery was simply marked on western maps, "Here be dragons."

Cruise passengers will find more diversity of cultures and countries in Asia than anywhere else in the world.

But today the Orient also has some of the world's most modern cities, high-tech metropolises such as Tokyo, Hong Kong, Kuala Lumpur, and Singapore.

A short distance from Malaysia's Port Kelang (formerly Port Sweetenham), two of the world's highest buildings, the twin Petronas Towers, rise eighty-eight

BALI
BANGKOK
HONG KONG
KUALA LUMPUR
MANILA
PHUKET
RANGOON
SAIGON
SHANGHAI
SINGAPORE
TAIPEI
TIANJIN
YOKOHAMA

Traditional opera has graced China's stages since the thirteenth century.

stories above Kuala Lumpur (Malay for "muddy estuary"). The awesome, towering structures are a futuristic fantasy, like something from Batman's Gotham City. And there is nothing quite like seeing Hong Kong's dramatic skyline from the deck of a ship moored at Ocean Center, with the giant high-rises built halfway up the steep mountain slopes of Hong Kong Island, or the colorful band of neon glowing along the harborfront at night.

But a passenger approaching a port by cruise ship can sense much of the ancient East that had such a strong pull on early voyagers. In Singapore, bum boats with magic eyes painted on their prows ply the world's busiest harbor between Boat Quay and cargo ships weighing at anchor. The world's largest sailing fleet, the wood-hulled, two-masted pinisi ships of the Bugis people, is moored at Jakarta harbor, Indonesia. Bugis people, once fearsome pirates, were the original "bogeymen" used to frighten naughty children. Traditional junks with their distinctive wooden hulls navigate through Hong Kong's bustling harbor, although now powered by diesel engines rather than sails. But on some magical days, a batwing junk in full sail—a tour boat taking some lucky visitors out on a sightseeing or dinner cruise—passes majestically past Hong Kong along one of the world's great waterways.

In Bangkok, strings of elephantine rice barges slowly drift down the Chao Phrya River while powerful long-tail boats race along like riverain hot-rodders, churning up the murky water. At the riverside Wat Arun (Temple of the Dawn), in the shadow of a 282-foot stone spire a happy fairground atmosphere prevails, with steaming outdoor food kitchens cooking spicy Thai noodles and shaded stands displaying a quaint collection of souvenirs. Across the river, the Grand Palace is a psychedelic wonderland of patterns and strange shapes, of mirrors, mythical creatures, gilded stupas, and the dazzling Emerald Buddha.

Tokyo's huge Meiji shrine is more subdued and natural, but impressive in an understated way. Shinto gates and temples are mainly of natural, unadorned wood with symbolic pieces of white paper folded like lightning bolts and huge sacred ropes hanging from entry gates. Priestesses in baggy, bright Orange Crush-colored skirts and monks in all white move through the grounds like wraiths. The boomboomboom of a drum echoes from within the temple, while supplicants buy paper prayers to tie to trees, where they sit like so many white butterflies.

Junks are a fading sight in the beautiful Hong Kong Harbour waters.

In many Asian cities, fine colonial buildings recall a romantic era, of ceiling fans and planters' chairs, and shady, colonnaded terraces where planters and district officers sipped chota pegs (single shots) of whiskey and soda. Deep in the vast jungles, British, Dutch, and Portuguese forts with ancient rusting cannons and thick, moss-covered walls molder in obscurity along remote muddy rivers.

Most Asian ports have a grand hotel steeped in history as a legacy of colonial days, when it was perhaps the only acceptable accommodation for foreign travelers. In their spaciousness, old-world tranquility, elegance, architecture, and sense of history, the grand hotels recall the age of leisurely journeys to faraway places when travel was for a select few. In those unhurried days, privileged voyagers brought servants bearing massive steamer trunks. Although most have adapted to the jet age, these venerable hostelries, remnants of the steamship and railway era, are worth a visit to soak in the atmosphere over tea in the lounge or a drink in the bar.

Hong Kong's Peninsula Hotel on what was once known as Fragrant Harbor had a storage room in its refrigeration plant in which ship passengers' furs could be kept on hot tropical days. Its vast, elegant lobby is still the gathering place for locals as well as tourists. Novelist Noel Coward penned the famous lines about mad dogs and Englishmen, and Hong Kong's Noon Day Gun while staying at the Pen. And Bangkok's venerable Oriental Hotel is redolent of the days when Conrad, Coward, Maugham, Kipling, and other tellers of tales-of-the-East stayed there. The outdoor Riverside Terrace overlooks

Singapore's annual Chingay Festival is a showcase for Chinese opera not to be missed by music lovers.

THE VISIONS OF

SINGAPORE

"My new colony thrives most rapidly," wrote Sir Thomas Stamford Raffles in 1819 shortly after he founded Singapore. Nearly two centuries later, the tiny island has grown into everything he envisioned and more: one of the world's largest ports, one of Asia's most important business centers, and a model of social symmetry where Chinese, Malays, and Indians live in harmony. Many visitors pack their bags with visions of incense and orchids, and that's exactly what they find in Singapore's ethnic enclaves—Chinatown, Little India, and Kampong Glam—where life seems to have changed little since the days of the British raj. ⚓

P*ort appeal rating*

Adventure	★★★★
Entertainment	★★★½
Romance	★★★½
Cuisine	★★★★½
Shopping	★★★★½

the broad Chao Phrya (River of Kings), the city's great aquatic highway. Singapore's Raffles, one of the most famous hotels of the East, still retains a little of its old character, despite now being a part of a huge, modern hotel.

The Manila Hotel, overlooking the broad sweep of palm-lined Manila Bay, has special meaning for Americans. The white stucco structure, designed in California-mission-style architecture, was a center of Manila social life for most of this century. The British-built Strand in Rangoon, Burma (now Yangon, Myanmar), with its grand, columned portico, has been restored to its old splendor. The teak-furnished inn with just thirty-two rooms features the turn-of-the-century Strand Grill, a wood-paneled bar with mahogany tables and chairs, marble tiled floor, and the Lobby Lounge with colonial cane furniture and twirling ceiling fans (now assisted by air conditioners). While in Saigon (officially Ho Chi Minh City), the French-colonial-style Continental is the preferred choice for character (as well as location),

although no longer the most luxurious hotel in the Vietnamese city. The Continental was a favorite meeting place of the expatriate community in both the French and American colonial eras. From the open-air terrace (then known as the Continental Shelf), expatriates and visitors looked out over the broad street at delicate young Vietnamese ladies in cone hats and in filmy ao dais (long silk dresses worn over pants) pedaling by on their bicycles. In Japan, the Fujiya in the Fuji-Hakone National Park, which opened in 1878 as the country's very first foreign-style hotel, is

Hong Kong's Wanchai waterfront glitters at dusk.

THE BECKONING OF

HONG KONG

With its glimmering skyline and ultramodern harbor, Hong Kong offers more of a glimpse of the future than a step into the past. But ancient days persist in parts of the former British colony, especially in the outlying islands where life seems to have changed little in the past hundred years. Some of Hong Kong's best seafood can be found in the outdoor cafés around Lamma, a favorite weekend retreat for local diners. Cheung Chau is an archetypal Chinese fishing village with a bustling waterfront and narrow alleys that reek of incense and dried fish. Lantau offers secluded beaches, hiking trails, and the world's largest bronze Buddha, a 250-ton statue that gazes serenely across the South China Sea. ⚓

Port appeal rating

Adventure	★★★★½
Entertainment	★★★★
Romance	★★★★½
Cuisine	★★★★½
Shopping	★★★★½

THE FACES OF

SAIGON

Its official name is Ho Chi Minh City, but tourists and residents alike still call it Saigon. By whatever name, Vietnam's largest city is one of the most vibrant ports in Southeast Asia, a metropolis of four million people that sprawls along the northern edge of the Mekong Delta. Dong Khoi Street is the heart of the city's tourism and nightlife zone, an area that stretches between Notre Dame Cathedral and the Saigon River. In bygone days, the street's sidewalk cafés were frequented by the movers and shakers of French society and famous writers like Somerset Maugham and Graham Greene. Today you can find some of Vietnam's best bargains along this stretch. ⚓

Port appeal rating

Adventure	★★★★½
Entertainment	★★★
Romance	★★★½
Cuisine	★★★½
Shopping	★★★★

The Saigon River snakes along the northern edge of the Mekong Delta.

now a serene mountain resort with a stunning vista of Mount Fuji, the nation's symbol.

Among Asia's greatest attractions is its wonderful food, different in every port. Those spices that lured the early European traders contribute to some of the world's oldest and most varied cuisines, and visitors here are in for incredible taste treats. Hong Kong competes with Taipei as the best city in the world for experiencing Chinese food, whether in a basic local restaurant or elegant hotel dining room. The multitude of migrants from all over China to this former British colony introduced every provincial variation of Chinese cooking to Hong Kong: Cantonese (the local cuisine), Chiu Chow, Peking, Shanghai, Szechuan, and even Hunan styles. New culinary experiences include Buddhist vegetarian restaurants and Hong Kong's own culinary creation, dim sum.

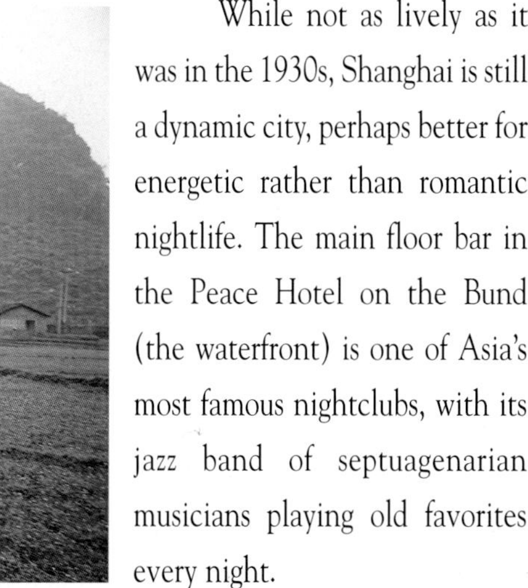

The limestone peaks around Guilin in southern China have inspired years of Chinese landscape painting.

Dining on these tasty snacks in a boisterous restaurant full of noise and life is a local tradition. And such Chinese delicacies as Peking duck and beggars' chicken are classic, not-to-be-missed dishes.

In Kuala Lumpur, visitors can sample piquant chili-prawns or some of Malaysia's fine coconut-flavored curries at street stalls or local restaurants. And Singapore's outdoor food stalls are among the world's best and most varied, with Malay, Indian, and Chinese food. On pleasant equatorial evenings, diners sit outdoors nibbling tasty satay (chicken, beef, or mutton shish kebabs served with a spicy-sweet peanut-chili sauce) and drinking cool coconut water, fruit juice, or local beer. But Thailand's highly distinctive cuisine is perhaps best enjoyed at an outdoor restaurant on the Chao Phrya River not far from the port. In this tropical setting, many diners will encounter the succulent tastes of tamarind, galangal, lemon grass, kaffir lime, and others for the first time. Dishes such as green and red curries, glass noodle salads, chicken in banana leaves, and sizzling hot seafood soups are all fantastic taste treats. In Jakarta, rijsttafel, literally "rice table," is an early example of fusion cuisine. The hungry Dutchman's version of the more ordinary Indonesian meal includes rice with numerous dishes of fish, meat, vegetables, and sambul (hot sauce) made from fresh chilies.

Balinese dance employs many different masks.

Couples looking for something more familiar—and romantic—in Asia's ports can have intimate, candlelit dinners, with the finest continental cuisine, served in scenic settings overlooking a river or harbor, or from a mountaintop. Home to many of the world's finest hotels, Asian cities have excellent classic western-style restaurants for intimate getaways. Shanghai, China's largest and most cosmopolitan city, was long known for its restaurants, especially legendary nightlife, theaters, dance halls, and bars.

While not as lively as it was in the 1930s, Shanghai is still a dynamic city, perhaps better for energetic rather than romantic nightlife. The main floor bar in the Peace Hotel on the Bund (the waterfront) is one of Asia's most famous nightclubs, with its jazz band of septuagenarian musicians playing old favorites every night.

Almost every Asian port offers a great variety of nightlife, from bars, restaurants, and clubs to disco (a western export to the East) and karaoke (an eastern export to the West). Tokyo has its Ginza, Shinjuku, and other districts while in Singapore, the action is at Boat Quay, an excellent rejuvenated riverside area that is now a bar and restaurant strip. Hong Kong's spirited nightspots are at Lan Kwai Fong, Wanchai (of *The World of Suzy Wong* fame), and Tsim Sha Tsui (right next to the cruise ship terminal).

A spirit of peace permeates modern-day Vietnam.

Equine images are popular in Taiwan.

A fireboat often welcomes cruise ships to Osaka.

And every country offers local cultural shows, with song, dance, and native costume. Early visitors to Bali marveled at the incredible classical barong or legon dances and the shadow or puppet performances of Ramayana and Mahabaratha Hindu epics. On this Island of Gods and Demons, perhaps the most spiritual, religious, and traditional enclave in the whole mysterious East, these dances and plays, which tell incredible stories of the ancient Hindu gods, are performed on moonlit nights in temple compounds.

In Japan, culture lovers can see Kabuki, Bunraku, and noh plays, which are like nothing they have ever experienced before. Distinctive, four-hundred-year-old Kabuki dramas—flamboyant song and dance shows with melodramatic plots—are bewildering, even bizarre to westerners, yet they are also captivating. The stage settings are colorful, the costumes extravagant, with elaborate kimonos, sashes, fans, and hairpieces. And the acting is highly stylized, with women (always played by male actors) plastered in thick white pancake makeup like that of circus clowns. While in ports with large Chinese communities such as Hong Kong, Singapore, Kuala Lumpur, Macau, or Taipei, visitors could happen on a lion or a dragon dance, with all of their excitement, color, and noise provided by clanging cymbals and exploding firecrackers.

Travelers are sure to find authentic Chinese culture at Hong Kong's Temple Street Night Market, where fortune-tellers read palms (and faces) to the backdrop cacophony of the wail of Cantonese opera performed right on the street, next to temporary stalls piled with modern goods, from bleeping clocks to casual clothes.

Japan is a feast for both the eye and palate.

Shopping is a major attraction in Asia's bustling cities. Early traders came on shopping (or rather, trading) expeditions, in search of silks and spices, tea, pearls, and opium. The only opium sold now is a brand of perfume, but all of the other wonderful treasures of the East are available in tiny, aromatic shops or grand shopping malls, along with the finest European designer wear. Asian cities such as Hong Kong, Singapore, and Seoul are consistently voted the world's best places for shopping for high-fashion items, and their deluxe megamalls are truly impressive emporiums. You'll find the latest in modern electrical appliances in Asia as well—especially in Tokyo. Major Japanese electronics manufacturers test-run their latest products in the specialty stores in the city's Akihabara district, known as denkigai (Electric City). Even non-shoppers can delight in the state-of-the-art merchandise that won't be available in the U.S. until about a year later. Products from the Spice Islands sold in many markets make great souvenirs. Pepper, cloves, nutmeg, chilies, peanuts, coconut, ginger, coriander, tamarind, lime, anise, cinnamon, candlenuts, lemon grass, and more all come in small packages, so when amateur chefs return home, they can re-create the new dishes that they sampled in Asia. Silk from China, Japan, or Thailand is also an evocative, easy-to-pack souvenir. Unique Thai silk, with its distinctive texture and bright sheen, is an especially good buy. The richly colored, shimmering material so emblematic of the fabled East is used for clothing, furniture, even for interior decoration.

Kyoto boasts more than sixteen hundred temples.

In Jakarta's Pasar Antik (antique bazaar), more than one hundred stalls overflow with incredible arts and crafts that reflect Indonesia's cultural diversity. Shops stock Irian Jayan carvings, Javanese puppets, Kalimantan blowguns, bronze Buddha heads from Yogyakarta, brass gamelan gongs from Sumatra, stringed musical instruments from Batavia. More exotic wares include intricate Batak calendars carved on water buffalo hipbones or ribs, buffalo-horn rice spoons from Java, and buffalo-skin wayang kulit, or shadow puppets, from Palembang, Sumatra. Colorful textiles such as South Sulawesi silks; woven ikat material from Toraja, Sulawesi; and Javanese batik make wonderful souvenirs. Batik, textiles waxed and dyed with elaborate designs, often by hand, is Indonesia's best-known folk art, sold in markets as well as department stores. Patterns can be floral, free-form, paisley, or Daliesque, and often feature animal motifs, imaginative geometric designs, or illustrations of local village life. And it seems that all of Bali is a crafts marketplace. Carved masks, silver jewelry, paintings

Even in Toyko the geisha culture survives amid modern western influences.

THE HAPPENINGS OF

YOKOHAMA

This port city in Tokyo Bay serves as the gateway to Tokyo. It's often hard for people to come to terms with the world's largest city with nearly thirty million people. But the Japanese capital is easy to tackle if you simply view the metropolis as a cluster of villages with distinct personalities. Ginza is the flashy, neon-shrouded shopping area where huge department stores compete with chic boutiques. Roppongi is a rowdy entertainment district filled with discos and nightclubs that rock around the clock. Shinjuku is an Asian Manhattan with its skyscrapers and business-suit brigades. Tsukiji is the pungent haunt of sailors and sushi sellers, a lively fish market that unfolds each morning before dawn. Asakusa basks in the Japanese past, a place of tranquil gardens and splendid temples. ⚓

P*ort appeal rating*

Adventure	★★★★½
Entertainment	★★★★
Romance	★★★
Cuisine	★★★★
Shopping	★★★★

THE TEXTURES OF

BALI

Waves of foreign influence have pounded the shores of Bali for thousands of years. Yet no matter what their strength, they are unable to erode the island's indigenous culture. The Balinese have created a fabulous world of dance, music, art, and weaving that reverberates through nearly every valley and village. Tourism has spurred a creative renaissance and a worldwide yearning for Balinese arts and crafts. In essence, Bali has reversed the process of commercial colonization. Rather than become a mirror image of western materialism, it has turned the tables and begun to imprint its culture upon the West. ⚓

Port appeal rating

Adventure	★★★★½
Entertainment	★★★★
Romance	★★★★½
Cuisine	★★★★
Shopping	★★★★

Rice terraces cover much of the Balinese hinterland around Ubud.

THE PEARLS OF

PHUKET

People searching for the perfect paradise island need look no further than Phuket, a magical destination that Thais call the Pearl of the South. Every beach is a study in perfection: fine white sand framed by coconut palms and turquoise water, a backdrop of jungle-shrouded hills, and a hammock perched at just the right angle for an afternoon snooze. The offshore reefs harbor a treasure trove of marine life and some of the best diving in Asia. And the local cuisine — especially the hot and spicy coconut curries — is simply to die for. It's easy to see why people come to Phuket for a week and decide to stay for years. ⚓

Port appeal rating

Adventure	★★★★
Entertainment	★★★
Romance	★★★★
Cuisine	★★★½
Shopping	★★★½

of primitive Balinese scenes, puppets, and of course, T-shirts are all sold on the streets.

Adventurous travelers looking beyond the shopping malls will find plenty of action in Bali, as well. Indonesia's rivers are great for white-water rafting, with Bali's Ayung and the more challenging Balian rivers especially accessible to overseas visitors. Moderately fit hikers can climb Mount Fuji, near Tokyo, staying overnight in huts on the mountain, to arrive at the peak at sunrise. Mount Kinabalu, in Sabah, Malaysia, the highest mountain in Southeast Asia at 13,455 feet, is more challenging, but can be climbed without mountaineering experience or equipment. Any fit person can make the ascent, although it is too arduous for the nonathletic.

Many lavishly decorated figures welcome visitors to the Wat Phra Kaew temple in Bangkok, Thailand.

Clear waters not far from Bangkok, Kuala Lumpur, Manila, Singapore, or Bali offer excellent snorkeling and scuba diving. Diving, now a major sport across Asia, is well organized for visitors. Diving operators and resorts in idyllic spots such as Puerto Gallera, just a few hours from Manila, provide all of the necessary equipment, and training as well. Boating of all kinds is also very popular in the region, and convenient, particularly in the Philippines and Thailand. Many divers now combine sailing and diving, using boats as floating hotels as well as diving platforms, in places such as Phuket, near Bangkok. More offbeat is kayaking in scenic Phuket, Thailand, or in Ha Long Bay near Hanoi, Vietnam. Operators run overnight trips or weeklong adventures, with English-speaking guides. Paddling the maneuverable but stable Sea Explorer kayaks, boaters explore the hundreds of small, weirdly shaped, karstic islands.

Southeast Asia's forests provide some rewarding bird and wildlife viewing, as well. The most unusual prey of all is the Komodo dragon, the world's largest lizard, on Komodo Island east of Bali. Mapmakers of old only guessed, but cruise visitors can personally attest that "Here be dragons."

A Malaysian taxi navigating through traffic reflects the culture's blend of old and new.

From Yokohama in the north to Singapore, Bali, and smaller ports in southern seas, Asia is a nirvana of fantastic adventures and sights. Cruising is relatively new to the area, but cruise passengers will discover a kaleidoscope of cultural experiences, a rich array of sights, sounds, scents, and tastes. Asia is as piquant as a chicken curry served by a roadside hawker, as lively as a lion dance, as mysterious and timeless as the jungle- covered stone ruins of Cambodia's world heritage site of Angkor Wat, and as modern as the latest computerized gadgets sold in sky-high department stores of Taipei and Tokyo. All the mysteries of the East that first astounded maritime explorers who sailed into these wondrous waters just a few centuries ago are open to today's travelers—and they don't have to climb the rigging to get there. ⚓

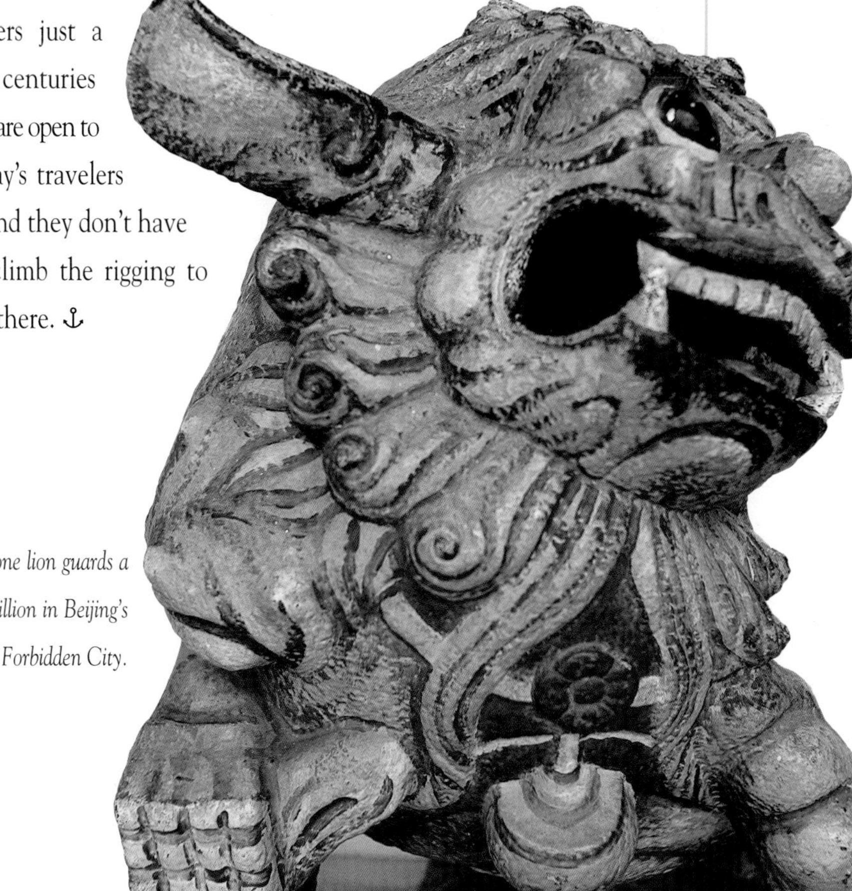

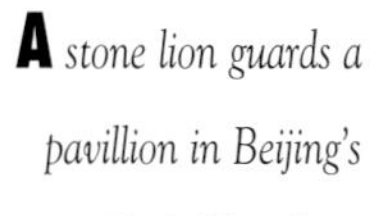

A stone lion guards a pavillion in Beijing's Forbidden City.

The many masks of Phuket hint at the magical allure of Thai culture.

Thai classical dance includes elaborate costumes inspired by the Ramayana epic.

THE TRADITIONS OF

BANGKOK

Royal tradition holds firm in Thailand, where the king remains the nation's most respected man and the queen one of its more revered women. The royal family dwells behind the whitewashed walls and beneath the golden domes of the Grand Palace in Bangkok. Much of the palace is open to visitors, a fascinating insight into Thailand's regal life over the past two hundred years. A great courtyard is surrounded by historic buildings, including the audience hall where kings doled out royal justice and a pavilion where they dismounted from their elephants. The ashes of bygone kings are housed within the adjacent Temple of Emerald Buddha, named for the country's single most important religious icon. ⚓

Port appeal rating

Adventure	★★★★½
Entertainment	★★★★
Romance	★★★½
Cuisine	★★★★½
Shopping	★★★★½

THE AMBIENCE OF

TAIPEI

Despite its modern hustle bustle, Taipei doesn't turn its back to ancient tradition. Taiwan's capital boasts some of the world's best Chinese art, including the countless treasures of the National Palace Museum, assembled by twenty emperors over a period of several hundred years. Traditional architecture reaches its zenith at the National Revolutionary Martyr's Shrine, a complex of Ming Dynasty buildings surrounded by tranquil gardens. Last but not least, don't miss the ancient culinary delights of Snake Alley with its exotic ambience and serpentine menus. ⚓

Port appeal rating

Adventure	★★★½
Entertainment	★★★
Romance	★★★
Cuisine	★★★★
Shopping	★★★½

Water-carved rock formations on Taiwan's northern tip form a natural stage for tai chi.

China's Great Wall, one of the seven wonders of the world, is near Beijing.

THE SIGHTS OF

TIANJIN

Situated at the confluence of five rivers, Tianjin is one of China's major ports and a thriving industrial center. It's also a gateway to the various delights of Beijing. The Chinese capital lies a mere eighty-five miles to the west, a convenient day trip for anyone interested in grandiose architecture or imperial history. Among the area's chief attractions are the Temple of Heaven, the Great Wall, the Summer Palace, and the vast Forbidden City. The local cuisine is also alluring, ranging from Peking duck roasted over a wood fire to an imperial banquet that can encompass as many as 170 dishes served over three days. ⚓

Port appeal rating

Adventure	★★★★
Entertainment	★★½
Romance	★★½
Cuisine	★★★½
Shopping	★★★

THE AWAKENING OF

SHANGHAI

China's largest city was the Pearl of Orient until the 1940s, when world war and revolution knocked Shanghai into a fifty-year phase of economic stagnation and cultural hibernation. The city's reawakening began in the late 1980s after sweeping economic reforms opened China's economy to the world. In little more than a decade, Shanghai has transformed itself from a drowsy backwater into one of Asia's most vibrant cities. A flurry of recent construction has produced China's finest art museum and the world's tallest skyscraper, but not at the expense of erasing the art deco elegance that made the old Shanghai such a fabulous port of call. ⚓

Port appeal rating

Adventure	★★★★
Entertainment	★★★
Romance	★★★½
Cuisine	★★★★
Shopping	★★★★

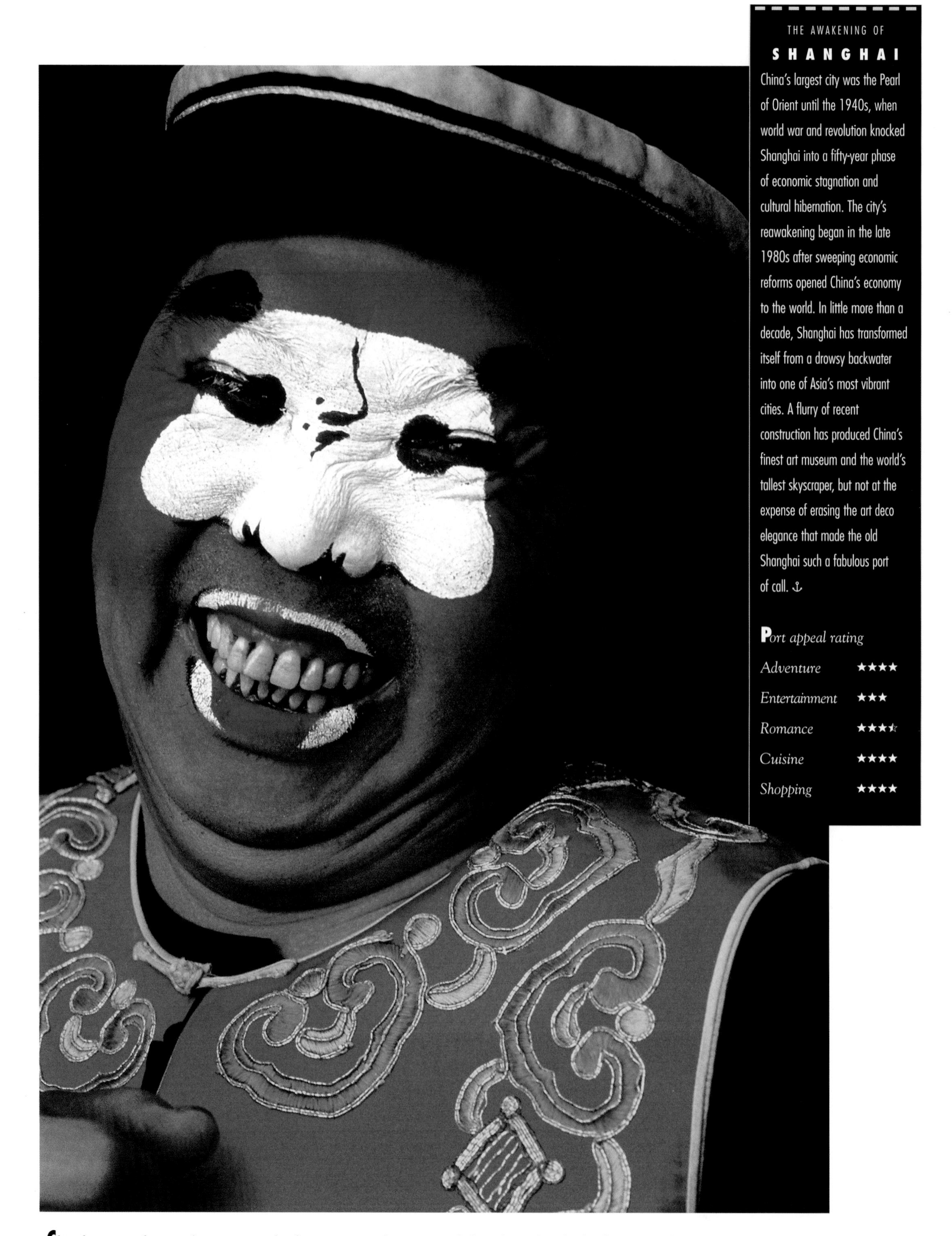

China has intrigued visitors for centuries as the ultimate in exotic destinations with the striking color of its local customs and arts.

The lost city of Angkor Wat hides amid the jungle of western Cambodia.

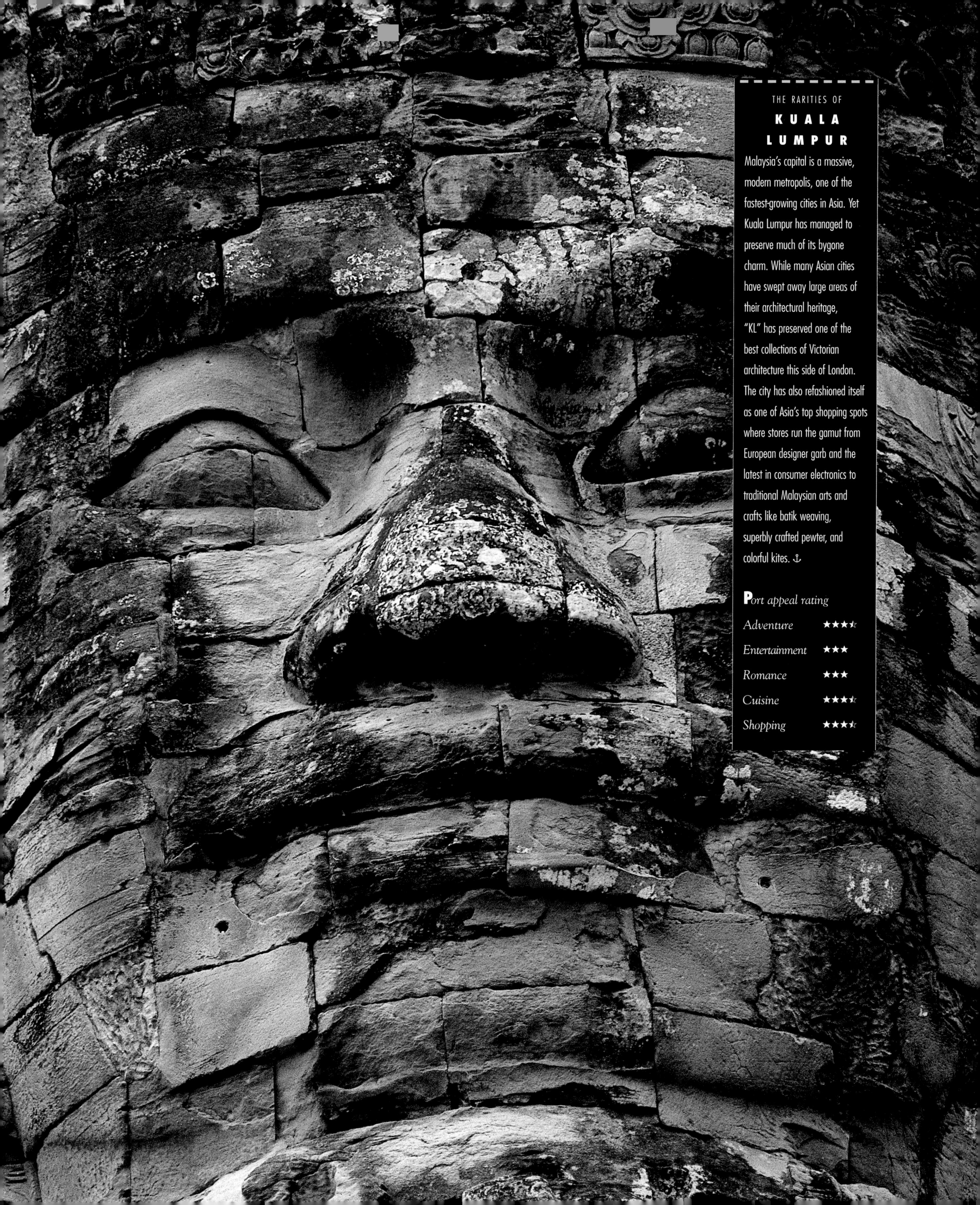

THE RARITIES OF

KUALA LUMPUR

Malaysia's capital is a massive, modern metropolis, one of the fastest-growing cities in Asia. Yet Kuala Lumpur has managed to preserve much of its bygone charm. While many Asian cities have swept away large areas of their architectural heritage, "KL" has preserved one of the best collections of Victorian architecture this side of London. The city has also refashioned itself as one of Asia's top shopping spots where stores run the gamut from European designer garb and the latest in consumer electronics to traditional Malaysian arts and crafts like batik weaving, superbly crafted pewter, and colorful kites. ⚓

Port appeal rating

Adventure	★★★½
Entertainment	★★★
Romance	★★★
Cuisine	★★★½
Shopping	★★★½

THE ECHOES OF

MANILA

Like so many Asian cities, the Filipino capital has grown by leaps and bounds in recent years. Yet bygone days persist behind the walls of Intramuros, the old colonial "city within a city" along the waterfront. Intramuros was founded in 1571 as Spain's primary bastion in the Far East. Four hundred years later it still has the feel of a medieval fortress with massive gates, moats, and cobblestone streets that echo with horse-drawn carriages. But Intramuros also has a contemporary spin. Many of Manila's finest restaurants, art galleries, and antique shops nestle behind the walls, and some of the ruins have been transformed into venues for outdoor theater, music, and dance. ⚓

Port appeal rating

Adventure	★★★★
Entertainment	★★★½
Romance	★★★
Cuisine	★★★½
Shopping	★★★½

THE POWER OF

RANGOON

Southeast Asia's most impressive religious shrine crowns a hilltop in southern Rangoon. Shwedagon Pagoda defies written description or photographs; you must cast your own eyes upon this magnificent structure and feel the celestial power that seeps from every stone and tile. The golden stupa—as tall as a thirty-story building—glistens in the noonday sun, visible from miles around. Arrayed around this central shrine are fifteen hundred gold and silver bells that tinkle even in the softest breeze and numerous altars where the faithful leave offerings of incense and food. Thousands of devotees appear each day, circling the stupa in clockwise fashion, their feet shuffling against the marble pavement as they recite their sacred mantras. ⚓

Port appeal rating

Adventure	★★★★½
Entertainment	★★★
Romance	★★★½
Cuisine	★★★
Shopping	★★★½

Rangoon and its sister city across the bay, Mandalay, Mynamar (Burma), are resplendent with temples.

The rich and colorful spirit of Manila is demonstrated in costume and dance.

BY LORRY HEVERLY

A·F·R·I·C·A

THE GREAT MYSTERIOUS LAND

IN THE FOOTSTEPS OF legendary explorers and adventurers, the great and mysterious African continent has always held a special place in the hearts of those who disembark upon its shores. Unforgettable African sunsets, calls of the wild in the bush, the mysteries of ancient wonders, and exotic Arabian walled cities merge throughout the Dark Continent, in places that awaken the senses and captivate one's soul.

Intrepid wanderers, drawn to Africa's enticing ports, embark on journeys to capture the essence and ambience of a timeless land. Here ancient wonders often collide with modern technology and are unwillingly thrust into the twentieth century.

Those who seek the spirit of Africa are courted by a different drummer. Their hearts beat to the sound of a bustling, early-morning bazaar. Their pulses quicken as a pride of lions brazenly stalks along the scorched savanna following the scent of a kill. They crave the primal rhythm of ancient kingdoms, a tribute to the gods and rulers whose legacies withstand the sands of time.

CANARY ISLANDS
CAPE TOWN
CASABLANCA
DURBAN
MADEIRA
MAHÉ
MOMBASA
TANGIER
TUNIS

Africa is meant to be savored. As travelers wander in childlike innocence through this enchanted land, all that is Africa will permeate their every pore. Senses come alive in the feel of red clay dust clinging to the skin, the earthy smells of a marketplace, or a joyful smile that conveys understanding between two people, worlds apart.

This spirited and eclectic collection of countries is a melting pot of intriguing cultures and influences. To capture the essence of an African experience is to cherish every morsel of its astounding abundance and to open the heart, mind, and soul for the adventure of a lifetime.

Castaway island outposts, like far-flung coins tossed into the sea, cling to the orbit of Africa's mainland. It is possible to retrace the paths of epic voyages along the ancient sea-trading routes to islands that glitter like emeralds in the azure Indian Ocean.

Here, sailing ships of long ago carried the bounty of exotic spices, mammoth tusks of ivory, and captured slaves. Crafty pirates lay patiently in wait to pillage and plunder unsuspecting merchant ships,

Masai people still inhabit a large part of Kenya's Great Rift Valley.

retreating with their prizes to a tropical hideaway in the Seychelles. Today, this cluster of breathtaking coral islands, a thousand miles off the Kenyan coast, lures cruisers to bask on sun-drenched, linen-white beaches while caressed by spice-swept breezes. Beachcombers could spend months exploring the many palm-fringed beaches of Mahé, staking a claim on their slice of paradise. However, the beaches are just a hint of the islands' allure. While waves lick the shores of a barefoot afternoon, an undersea wonderland awaits discovery.

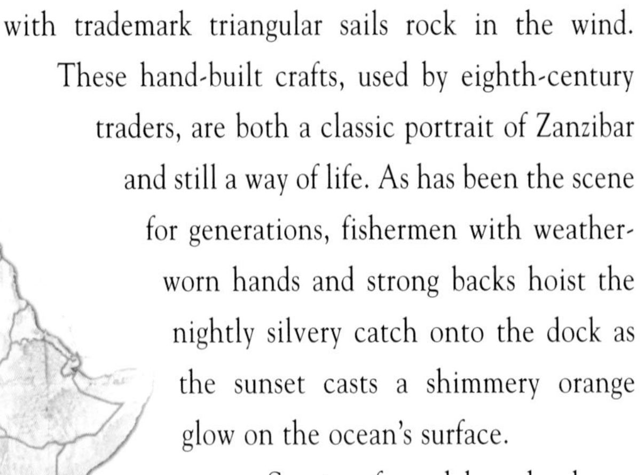

Dancing in the ocean's gentle currents, vibrant coral reefs create a fairytale landscape of surreal pinnacles and enchanting grottos. Here, a kaleidoscope of exotic rainbow-colored inhabitants shares liquid space with adventureous snorkelers and scuba divers.

Nearby islands take on a captivating charisma of their own. The speed of a rickety ox-drawn cart—a fun way to explore the tiny outpost island of La Digue—is how villagers measure the pace of life. Endearing couples strolling hand in hand on Praslin Island are amused as they pass the erotic double-nut-shaped fruit that grows from the rare coco-de-mer palm. Bird-watchers enter a wing-and-feather heaven on Bird Island, home to millions of nesting terns. Offshore, the fight is on as fishermen match wits, looking for just the right lure to snag a world-class marlin, sailfish, or bonefish.

Zanzibar. Just the name of this spice island conjures up exotic images. Here Africa meets Arabia off the Tanzanian coast. Elaborate palaces, Persian baths, and opulent dwellings for harems are traces of the island's wealth and splendor from magical days under the rule of the sultans of Oman. Tales of this faraway land are captured in stories of *Arabian Nights* and in the adventures of Sinbad the sailor.

Lining the harbor, Arab dhows with trademark triangular sails rock in the wind. These hand-built crafts, used by eighth-century traders, are both a classic portrait of Zanzibar and still a way of life. As has been the scene for generations, fishermen with weather-worn hands and strong backs hoist the nightly silvery catch onto the dock as the sunset casts a shimmery orange glow on the ocean's surface.

Scents of sandalwood, cloves, vanilla, and eastern spices perfume the breezes in the local market. The cobblestone streets of old Zanzibar's Stone Town are a jumbled collection of shops and cafés nestled between intricately carved, brass-studded doors of mosques and palaces. A haunting sound from the mosques calls the faithful to pause for daily prayer. One could spend forever exploring the nooks and crannies of this and the many intriguing islands of the Indian Ocean, but adventures of a wild kind await voyagers on East Africa's mainland.

For centuries Mombasa was a trading center for Persians, Turks, Indians, Portuguese, and the British. A fascinating potpourri of forts, temples, and mosques remind visitors of its many occupants. Prosperous fifteenth-century merchants once walked the streets in cloths of gold. This ancient Kenyan frontier is not only a port of plenty—inviting beaches, historic spots, seaside restaurants, and jumping nightlife—but also the gateway to the heartland of great wildlife expeditions.

Ceremonial masks suggest the artistry and colorful display of African cultures that await visitors.

Guests of the Windsor Golf and Country Club near Nairobi discover an oasis of the western world on the plains of East Africa.

The Seychelles comprise eighty-six islands that lie about halfway between Kenya and India.

THE LURE OF

MAHÉ

Prior to the arrival of the first human beings in the seventeenth century, the Seychelles Islands were dominated by huge, hulking prehistoric animals. We're not talking dinosaurs here, but the giant tortoise, which is prominently displayed on the country's coat of arms. Mahé is one of the only two places on the planet (the other is the Galapagos Islands) where the mammoth reptiles are found. They were once widespread throughout the archipelago but overhunting during early colonial days greatly restricted their range to a single island, secluded Aldabra, where several thousand roam today. Tortoises have been transplanted to more central isles like Curieuse, Bird, and Cousin, closer to the cruise port at Mahé. ⚓

Port appeal rating

Adventure	★★★★½
Entertainment	★★★
Romance	★★★★★
Cuisine	★★★½
Shopping	★★½

THE SPIRIT OF

MOMBASA

Kenya's primary port is the spiritual capital of the Swahili culture that once dominated the East African coast from southern Somalia to northern Mozambique. Much of the city's character derives from this influence: women in full-length black dresses, old men in flowing white robes and embroidered skullcaps, the proliferation of mosques, and the narrow streets of the Old Town. Arabic arches and minarets cast exotic shadows across the northern harbor, one of the last bastions of the lateen-sailed dhow. Fort Jesus may have been built by the Portuguese, but it was later captured by the Swahilis and transformed into their power base for this region. ⚓

Port appeal rating

Adventure	★★★★½
Entertainment	★★★
Romance	★★★★
Cuisine	★★★
Shopping	★★★

Vast herds move back and forth between Kenya's Maasai Mara and the adjacent Serengeti Plains.

The word *safari* in Swahili simply means to travel or make a journey, but here a simple journey is unlike any other on earth. Far away from hectic paces and busy lifestyles are places where wild creatures run free and one finds happiness in the simple things.

A towering giraffe nibbles leaves from the treetops. Zebras rub shoulders while enjoying a refreshing drink from the stream. A submerged hippopotamus occasionally cracks a widemouthed yawn, while clusters of graceful gazelles dance across the field.

Game trackers with eyes as sharp as the African fish eagle follow the spoor of elusive wildlife. In the bush a sighting could always be around the next dusty bend. Everyone is on full alert, cameras uncapped and binoculars ready. Long sleeves and hats deflect the powerful African sun. Suddenly from the sidelines a pair of Thomson's gazelles dash across the trail in front of the jeep. Like a game of hide-and-seek, lions, elephants, cheetahs, giraffes, and zebras await discovery in the rugged wilderness and vast open spaces.

Tsavo National Park is home to the legendary red elephants, the huge tuskers aptly named for their habit of rolling in the rich clay mud. Two million wildebeests in a dusty thundering herd charge across the Serengeti Plain on their annual migration. The wonders of nature abound here in game reserves like the Maasai Mara, Samburu, and Amboseli.

Kenya is rich in geographic diversity. Flat savanna plains, highlands covered in coffee plantations along the equator, lake regions—

The Okavango Delta of Botswana are one of southern Africa's best-kept wildlife secrets.

THE WILDS OF

DURBAN

Durban is the gateway to some of Africa's best animal viewing, with its string of national parks and game reserves along the northern coast of Natal. Umfolozi is the largest reserve, a great breeding ground for rhinos (both black and white) and the only Natal park with a large lion population. Mkuzi and Hluhluwe (pronounced shlew-shlew-wee) are famed for their water holes, where you can watch wildlife from the safety of camouflaged hides. Ndumu is the best place to spot hippos and cheetahs. Besides these government-run reserves, a number of private game parks provide luxury accommodations and personalized safaris. ⚓

Port appeal rating

Adventure	★★★★
Entertainment	★★★½
Romance	★★★
Cuisine	★★★
Shopping	★★★

THE VIEWS OF

CAPE TOWN

Table Mountain hovers three thousand feet above Cape Town, a dramatic backdrop to the harbor and downtown area, as well as a huge nature preserve within minutes of the central city. Most people scale the mountain via the aerial tramway. But you can also walk up and down, via a trail that leads through Platteklip Gorge. The downward journey takes about an hour, the upward hike about three hours. The tableland at the summit has numerous trails. You probably won't see much wildlife, but there are plenty of exotic plants and stunning viewpoints that overlook Clifton, Camps Bay, and other beaches. For another stunning view of Cape Town, especially at night, you can take a taxi to the summit of Signal Hill. ⚓

Port appeal rating

Adventure	★★★★½
Entertainment	★★★½
Romance	★★★½
Cuisine	★★★½
Shopping	★★★½

home to flocks of flamingos, the Great Rift Valley, and magnificent Mount Kenya—are all a part of this country's allure.

In earthy mediums of reeds, wood, cloth, and clay, artisans and craftspeople create objects of desire for last-minute souvenir shoppers. Local markets explode in riotous color. Turbaned women share a laugh and the latest gossip, weaving intricate baskets from reeds and grasses. Young boys with nimble fingers carve wooden giraffes, elephants, and zebras that stand guard alongside bolts of gaily printed kangas, sarongs worn by local women.

The historic Vineyard Hotel graces a Cape Town suburb called Newlands.

Kind yet persistent merchants armed with enticing goods trail travelers at a respectful distance, shouting prices that are just too good to be turned down. In a final wave of bartering and exchanges, happy shoppers clutch material memories of an unforgettable East African adventure before they return to the ship.

The Cape of Good Hope is steeped in tales and superstition. Once a landmark for India-bound Portuguese explorers, the towering Table Mountains could be spotted a hundred miles out to sea. Long ago, Arab and Phoenician sailors superstitiously believed that the mountain held a magnetic force that would draw a ship to its doom.

Cape Town, with its dramatic mountain backdrop, is a distinctive-looking port with a sweeping bay and natural harbor. Here two great oceans, the Atlantic and the Indian, merge to become one. South Africa's wealth of farmlands, and diamond and gold mines, enticed early Dutch and British to establish its oldest settlement.

Cape Peninsula unfolds in quaint fishing villages, seaside towns, and idyllic beaches. Along a coastal drive, waves crash in a white spray against the rocky shore. Jackass penguins in tuxedo suits huddle together to ward off the strong winds at Boulder Beach. Road signs in Afrikaans and English warn visitors not to feed the baboons. Troops of these menacing primates roam the peninsula and hang out in parking lots hoping to snatch a free lunch from unsuspecting picnickers.

Sweeping vineyard estates rest in the shadow of mountainsides, in parcels of a patchwork carpet. Cruisers with discerning palates will savor a visit to the world-class wine regions of Paarl and Stellenbosch. Wooden kegs of fermenting grapes, stacked to the ceiling of a wine cellar, remain untouched until their time has come. Some of the world's finest reserves are on hand for tasting and buying pleasures.

Cape Town flaunts wonderful beaches and rugged coastal scenery.

Durban, a lush subtropical port, boasts a large Indian population. Thus it isn't surprising to see rick-shaw drivers parading tourists through the streets to explore Hindu temples, find restaurants featuring spicy curries, or discover the best shopping at the Indian market. Skyscrapers and sophisticated year-round resorts share the waterfront with golden beaches that are perfect for surfers and sun worshipers.

Beyond the city limits, Zulu tribesmen keep proud traditions alive with displays of energetic dances and arts and crafts at the enormous Kraal of King Shaka. The great battlefields where Zulu warriors, Boers, and the British struggled for control remind visitors of South Africa's often turbulent past. However, like the mythical Phoenix from the ashes, the country continues to rise above its differences, reshaping history into the new millennium.

Romantic French and exotic Arabian flavors flow together in northern Africa's ports of Morocco and Tunisia. A ramble through the catacombs of a Casbah, bazaar, or medina overwhelms the senses. Men sip tiny cups of thick Arabic coffee, catching up on the day's events at sidewalk cafés. Some partake in a game of chess, while others linger over a newspaper. Many women in the old quarters dress from head to toe in black, covering the lower half of their faces in cloth draping.

Nestled in hidden alleyways, endless rows of shops offer a dizzying array of choices. Shiny brass plates, tea servers, and knickknacks

Giraffes munch their way across a Kenyan landscape.

THE DRAMA OF

CANARY ISLANDS

The starkly handsome Canary Islands lie off the west coast of Morocco, but for more than five hundred years they have been part of Spain, a jumping-off point for voyages of exploration to the New World and now one of Europe's most popular "sun and sand" destinations. The seven major islands all have their own special character. Grand Canary is the most Iberian, especially the bustling port city of Las Palmas. Lanzarote offers dramatic volcanic landscapes. Tenerife, the largest island, is renowned for its lush gardens and verdant farm fields. Hierro, the smallest isle, is an unpretentious place where life seems to have changed very little in the last hundred years.
The Canaries are one of the few places that live up to the hackneyed phrase, something for everyone. ⚓

Port appeal rating

Adventure	★★★½
Entertainment	★★½
Romance	★★★½
Cuisine	★★★
Shopping	★★½

An excursion aboard a dromedary camel is a highlight of a visit to volcanic hillsides of Lanzarote in the Canary Islands.

Brightly painted fishing boats color the beaches and bays of Maderia.

THE GUSTO OF

MADEIRA

Roughly 350 miles off the African coast, the Portuguese islands of Madeira offer a rare combination of adventure, sport, and gourmet dining. Surrounded by the deep blue Atlantic, Madeira boasts all sorts of water sports including sailing, diving, surfing, and deep-sea fishing. Madeira wine has been famous for centuries, originally fortified with brandy for the long sea voyage to Europe, but now offered in a number of varieties. The island wine goes great with local edibles like black swordfish, sweet potato bread, codfish pie, and lamb stew. ⚓

Port appeal rating

Adventure	★★★★
Entertainment	★★½
Romance	★★★★
Cuisine	★★★½
Shopping	★★★½

reflect the rays of the midday sun. Bangles and baubles in gold, handcrafted silver, and semiprecious stones glint and glimmer in eye-catching displays. Goods of kid leather—wallets, belts, jackets, and pillows—clutter makeshift shelves. Measured by the knots per square inch, thousands of rugs, from dowry rugs in muted earth tones to exquisite prayer rugs of pure silk, are prize acquisitions here.

Shopping becomes a fine art of negotiation since nothing has a price and everything is for sale. Good-humored merchants offer shoppers the world. "Just come into my shop and look, only look" . . . "You are the first blue-eyes I see today" . . . "Please come in for a cup of mint tea.". . . Another merchant jokes, "I will give you many camels for her." It is this sort of banter and barter that turns a shopping excursion into an adventure.

Archaeological buffs will appreciate Tunisia's impressive collection of Phoenician and Roman ruins. Carthage, founded by the Phoenicians in 814 B.C.E., was the seat of power for over seven hundred years and the center of Christianity in Africa. Roman baths, the amphitheater, and the national museum offer a stunning assembly of Punic and Roman artifacts. The magnificent Roman amphitheater in El Jem, stranded remotely in the Sahara Desert, rivals the Colosseum.

The sand-swept reaches of the great Sahara Desert cover over half of Tunisia. Camel caravans once plowed along ancient desert trading routes, their pack animals heavily laden with silks, spices, silver, and jewels. Camels still drift through the Sahara along miles of endless desert landscape of golden rolling dunes, rugged mountains, and cracked salt lakes. On- and off-road tracks lead to tiny outcroppings of lush green oases fed by underground springs. Here, in remote villages, Berbers welcome parched and sand-laden travelers with refreshment and shade.

Against the blistering daylight and frozen desert nights of an inhospitable landscape, the Berbers have built ingenious dwellings like the subterranean city of MatMata where residents live underground. The only sign that one has reached the town is a collection of TV antennas jutting out of the sand. Hotels, restaurants, and shops are tucked away in dark, cavelike recesses at this troglodytic settlement.

Africa is a medley of mosque and minaret skylines in the north, a cluster of idyllic islands in the Indian Ocean, wild places where adventure-seeking souls head into the bush in the east, and the dramatic rugged coastlines, mountains, and vineyards around the great Cape in the southern stretches.

Once the destination of only the most daring and bold, the rich and well-traveled, today Africa welcomes thousands of visitors each year. One cannot help but fall deeply under the spell of this enchanted land. It is the voodoo chant of a witch doctor, the mystical beat of a native drum, and the enchantment of the Dark Continent's magical embrace that beckon the heart to return. Sailing away from this majestic region, one will concede that Africa has a way of stirring the soul and creating memories that can be found nowhere else on our great planet. ⚓

Luxor's massive Karnak Temple complex spreads across sixty acres.

Overleaf: The mighty Zambezi River tumbles 343 feet into a narrow gorge at Victoria Falls.

The mosques of Morocco are architectural delights as well as spiritual sanctuaries.

THE ALLURE OF

CASABLANCA

Don't expect to find Humphrey Bogart and Ingrid Bergman hanging out at Rick's Café—Casablanca has never really lived up to its celluloid image. As Morocco's largest business center and port, this is a hard-working city with its mind firmly set on the future rather than the nostalgic past. However, it's a marvelous gateway to the attractions of inland Morocco: the bustling markets and mosques of Marrakech, the ancient medina of Fez, the stark beauty of the High Atlas, and the oasis towns of the western Sahara. ⚓

Port appeal rating

Adventure	★★★★
Entertainment	★★★
Romance	★★★½
Cuisine	★★★½
Shopping	★★★

THE SPICE OF
TUNIS

After more than a hundred years of French colonial rule and several thousand years of close contact with the Roman and Greek civilizations, North African cuisine reaches a savory summit in Tunisia. Nearly every meal features some sort of couscous, a tasty semolina grain that can be topped with chicken, lamb, or vegetable sauce. Shish kebabs come in all shapes and sizes, not to be confused with kefta, which is actually ground meat cooked on a skewer. If you're in the mood for a salad try chakchouka with its subtle spices and finely diced onions, cucumbers, and tomatoes. Another good starter is harira soup, a wonderful blend of chicken and chickpeas. Speaking of chickpeas, don't forget the hummus. ⚓

Port appeal rating

Adventure	★★★★
Entertainment	★★½
Romance	★★★½
Cuisine	★★★
Shopping	★★★

The shops and stalls of Tunis present opportunities of acquisition from all sides.

The colorful dress of a water carrier reflects the liveliness of Moroccan city streets.

THE MYSTERIES OF

TANGIER

Tangier is one of the world's truly exotic cities, a meeting place of Europe and Africa on the threshold of the Strait of Gibraltar. The city was a Spanish enclave until 1956, when it was ceded to newly independent Morocco, but it retains much of its Iberian flare. In recent years it's become a bastion of alternative lifestyles practiced by a large population of expatriate Europeans. Tangier has everything you might expect of a North African city: a magnificent sultan's palace, a powerful hilltop fortress called the Casbah, a dark and sometimes dangerous bazaar, as well as snake charmers, camel herders, carpet sellers, and wandering troubadours. ⚓

Port appeal rating

Adventure	★★★★
Entertainment	★★★
Romance	★★★½
Cuisine	★★★½
Shopping	★★★½

Seabourn Sun *steams through the Suez Canal, which revolutionized sea travel between Europe and Asia after it was built in the 1860s.*

BY SHIRLEY STRESHINSKY

M·I·D·D·L·E E·A·S·T

ANCIENT AND EXOTIC

SOMETIME DURING THE NIGHT, the ship had sailed into the Aegean Sea and that part of the world called the Middle East, which reaches from the Mediterranean and the Aegean across the map to the very edge of India. By first light, the view from the top deck revealed the Dardanelles, the thirty-eight-mile-long strait that connects the Aegean to the Sea of Marmara and thence through the strait of Bosporus into the Black Sea, those waterways that separate the Occident from the Orient, Europe from Asia. Over the course of history, the Turkish straits have been of enormous strategic importance as world powers struggled to control them in order to establish supremacy over the Middle East. This water route—an object of conquest throughout history—is at once fascinating and confounding, a place where the cry of the muezzin, summoning the faithful to prayer, floats on the warm air, where the tension of history and mythology is almost palpable.

In ancient times the Dardanelles was called the Hellespont; Leander drowned trying to swim from one shore to the other to meet his lover, Hero, a priestess of Aphrodite. The Gallipoli Peninsula slides by in the early morning light, a poignant reminder of the opening salvos of the twentieth century and World War I. After crossing the Sea of Marmara on a bright blue day, the ship moves into the Bosporus, a narrow fifteen-mile-long channel that leads into the Black Sea, European Turkey on one side, Asian Turkey on the other. Cruise ships slide beneath the Sultan Mehmet Bridge. Along each side of the winding waterway are ancient fortifications, here and there a mosque or a palace, and elegant summer homes so close that people can be seen inside.

The voyage continues across the Black Sea, tracing the Crimean coast, to tie up fast alongside the palm-lined promenade of the Ukranian resort town of Yalta, where history has a habit of unfolding. On school holidays, the promenade will likely be alive with schoolchildren, little boys waving postcards at the disembarking passengers and shouting "One dollar! One dollar!" In fractured English, the boys offer to serve as private tour guides. Their sweet enthusiasm washes up and over the old streets, to the moldering Russian Orthodox church, where old women—babushkas or

ASHDOD
BOMBAY
ISTANBUL
IZMIR

A *wandering sadu (holy man) provides much of the local color greeting wayfarers in the streets of Kathmandu.*

Istanbul boasts one of the world's most exotic skylines.

grandmothers, as the boys call them—laboriously climb the stairs and disappear into the dim recesses of the old building, once again open for prayer after a long, Communist-era interruption.

Past the holiday houses—a warm climate and sea bathing have long made Yalta a popular vacation resort—is a lively, if spartan, indoor market. Here housewives engage in intense discussions with butchers and bakers, young mothers pushing baby carriages stop to visit, while on the park benches of the promenade, older men with colorful medals on their jackets turn their faces to the sun. The young tour guides with their impish smiles and angelic faces make it easy to approach other local people, who often are eager to talk.

The Church of the Sepulcher in Jerusalem, an important stop for Christian believers, is adorned with numerous sacred artworks.

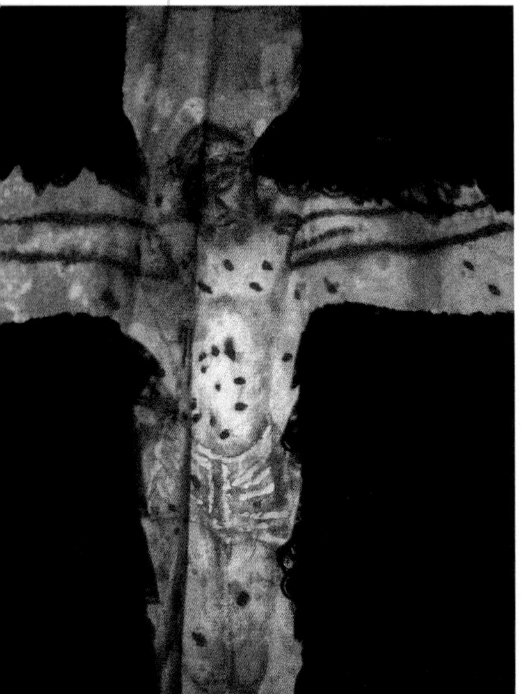

Tree-shaded paths in a great park lead to the White Palace, once the summer home of Czar Nicholas. In the palace's beautiful Great Hall, Roosevelt, Churchill, and Stalin came together half a century ago to decide the fate of postwar Europe. It is a dignified room, solemn and silent, as befits such an enormous moment in history. But outside, in the park, a variety of local people gather to entertain tourists for a few kopecks: a little girl in a pristine white dress and a matching giant bow in her hair plays the violin, and the classical notes seem to evaporate into the tall trees; a man puts his trained monkey through a series of tricks; babushkas spread out their hand-knit hats and scarves for sale, while others offer jars of caviar. Scattered through the elegant gardens are former Soviet soldiers selling off bits of their uniforms—belts, watches, hats—souvenirs of another time, another historical interlude.

If Yalta is subdued, Istanbul is a perfect tonic of a city, bursting with energy and a delicious sense of intrigue: altogether modern, perfectly Aladdin, with its minarets and mosques, Byzantine cathedrals, sultans and harem quarters, and the vast covered maze that is the Grand Bazaar. The bazaar holds intricately woven carpets, trinkets of beaten brass, air filled with the scent of steeped teas and exotic spices. It is all sound and smoke and splendor: the Blue Mosque, the Hagia Sophia, Topkapi with its great diamond, the Basilica Cistern.

On any brilliant spring morning, with seed motes and the promise of adventure floating on the sunlit air, the resourceful traveler can make the most of eight hours on shore by hiring a car and a guide. That way, the morning hours can be spent off the beaten track, perhaps joining the throngs of students who gather at the Kariye Museum to study what may be early Christendom's most passionate mosaics. After the Hagia Sophia, the Kariye—an early Christian church, and rather small—is the most important Byzantine building in Istanbul and one of the oldest, dating to the sixth century. Its walls are filled with mosaic art that pulses with life and feeling—the Virgin Mary as a baby taking her first steps, urged on by her doting parents; Mary looking adoringly at her own son. Daylight filters into the ancient building from skylights above, illuminating the very tender, human faces of saints, created out of bits of mosaic made centuries ago by unknown but hugely talented artists.

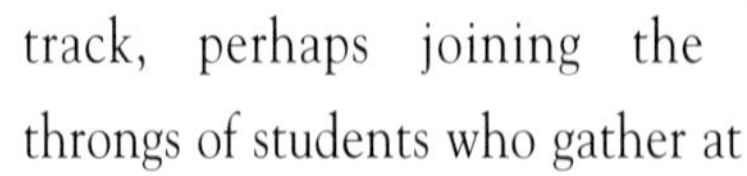

An elaborate shoe-shine box suggests one of the many offerings of the Grand Bazaar in Istanbul.

After so emotional an interlude, a visitor can pause in the tree-shaded square in front of the ancient church for Turkish coffee—thick and black—and one of the pretzel-like morsels sold from painted carts. In Istanbul, the past and present coexist with a sparkling elan. For an hour or so of fantasy relief, the traveler can wander through the opulent

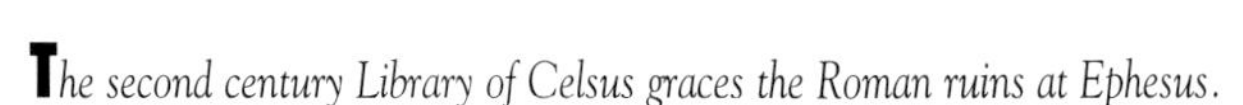

The second century Library of Celsus graces the Roman ruins at Ephesus.

THE RICHES OF

IZMIR

Izmir is a great place to sample Turkish cuisine. Drawing inspiration from Asia, Africa, and Europe, the native cuisine is the richest in the entire Middle East, a blend of roast meats, fresh fish, stuffed vegetables, mixed salad, pilaf rice, and truly scrumptious desserts. The lamb kebab in its various manifestations is the abiding trademark of Turkish cuisine. But don't forget other taste treats like meze (mixed hors d'oeuvres that might include caviar), biber dolmasi (stuffed green peppers), tarhana corbasi (tomato yogurt soup), and imam bayildi (eggplant with onions and parsley). ⚓

Port appeal rating

Adventure	★★★★
Entertainment	★★½
Romance	★★★½
Cuisine	★★★
Shopping	★★★

Visitors to Odessa can enjoy the lively color of traditional Ukrainian folk dances.

THE SPAN OF

ISTANBUL

Istanbul is one of the world's most beguiling cities and the only major metropolitan area that spans two continents. Founded more than sixteen hundred years ago during the waning days of the Roman Empire, the city grew into the capital of two of the greatest cultures that history has ever produced: the Byzantine and the Ottoman. Today's Istanbul, which glistens beside the dark blue Bosporus, is a product of both of these civilizations and the vibrancy of modern Turkey. Despite a plethora of historic sights—like the incomparable Hagia Sophia—the real attraction of Istanbul is the helter-skelter pace of local life as played out in the city's bustling markets, parks, and avenues. ⚓

Port appeal rating

Adventure	★★★★½
Entertainment	★★★★
Romance	★★★★½
Cuisine	★★★★
Shopping	★★★★½

harem quarters in Topkapi palace, home of the Ottoman sultans for four centuries, then pause for a cool drink in the leafy garden courtyard of a renovated hotel in the shadow of the Hagia Sophia.

The Turkish port city of Izmir, once known as Smyrna, goes straight to the throbbing heart of history. Namedropping knows no loftier heights: Homer was here; so was Alexander the Great and Marcus Aurelius, not to forget Süleyman the Magnificent and Tamerlane—none of whom would recognize today's Izmir, which is modern and bustling. But they would feel at home in nearby Ephesus, as in the biblical book called Ephesians, where Paul started his fabled address with, "To the saints which are at Ephesus. . ." and John wrote his Gospel, and the Virgin Mary is said to have lived out her life. It is one of the world's most astonishing ruins.

Old and new faces lend charm to Odessa, the Ukraine's leading Black Sea port.

Walking through the sprawling archaeological site on a sun-drenched day, bottle of water in hand, the late-twentieth-century visitor gets a surprisingly intimate glimpse of life in ancient times, from a fourth-century drainage system installed on a marble street to the library of Celsus built in A.D. 135, with its beautiful facade soaring fifty-three feet, to the brothels with their hidden passages (which might have provoked Paul's warnings, in Ephesians 5, against "fornication" and "whoremongers").

At the height of the summer season, the tour groups will melt one into another, until the ancient rock-strewn streets are thronged, this time with humans in shorts and ball caps and running shoes, and guides shouting in a dozen different languages to make themselves heard. At any moment, this cacophony could suddenly be stilled by the strains of the hymn "Amazing Grace" sung by a visiting youth choir sitting high up in the Great Theater, built in A.D. 54 to seat twenty-four thousand. The sweetness of the music swells and echoes out and over the ruins. A hush settles over the crowds, and for a few rare, shining moments, a feeling of glory returns to the great amphitheater.

Oh, Jerusalem! Heartland of three of the world's great religions —Islam, Christianity, and Judaism—and a place that lives in the minds and hearts of many more than will ever see it. On a moonlit night, the city glows a luminous white, like an alabaster dream. (Tending this particular dream is one small but revealing detail: a local law that requires all new structures to be built of one of three kinds of white limestone.) Within the Old City—one splendid square kilometer within walls built by the peripatetic Süleyman the Magnificent from 1537 to 1542—are some of the world's most holy sites: the Church of the Holy Sepulchre, believed to be the site of Jesus' crucifixion, burial, and resurrection; the Dome of the Rock, which covers a mass of gray stone held to the earth, Moslems believe, by the angel Gabriel on the night the prophet Muhammad ascended from it to heaven on his white horse, El Burak. There, too, is the Wailing Wall where for twenty centuries Jews have come to pray and mourn, the men to the left side and the women to the right. The ancient stones have been polished to a sheen by thousands of years of caresses, and tiny scraps of paper holding prayers are stuffed into every reachable crevice. Above it all is the echo of bells ringing in churches, of calls from minarets, of the cry of the shofar from synagogues, all summoning the faithful to prayer.

Icons are among the numerous emblems of Russian Orthodoxy and examples of the fine artistry to be enjoyed by passengers docked in Odessa.

A voyage through the Suez Canal into the Red Sea and then across the Arabian Sea ends just outside the far edge of the Middle East at the port city now called Mumbai—it was Bombay during the reign of

Istanbul's landmarks include the Blue Mosque (foreground) and the Hagia Sophia.

Historic buildings grace some of the more remote corners of Israel, accessible to the hardy traveler.

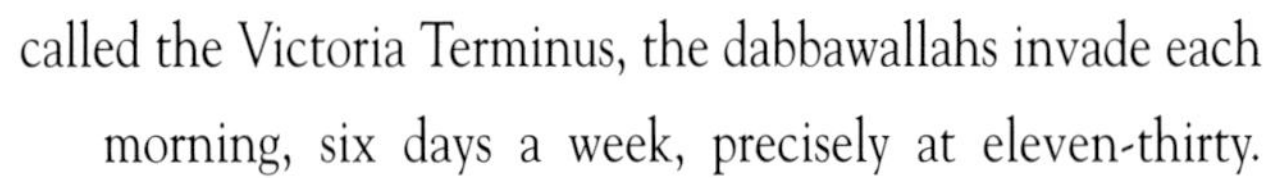

the British raj. (The local government has reinstated the name the original fisherfolk gave it, after a Hindu goddess named Mumbadevi.) British passengers disembarked near the triumphal arch called the Gateway to India, with a promenade that looks not so different today than when "burra sahibs" and "memsahibs" came ashore in the early decades of the twentieth century.

In India, like few other places on earth, are vestiges of life as it was lived in other centuries. The streets swirl with brilliant, fluorescent colors—cerise and jade green and saffron yellow, aglitter with silver and gold. Women wear graceful saris, even as they work in the fields; men wrap themselves in dhotis or wear white pajamas, often with the familiar Gandhi hat. Smoke drifts through the streets from braziers fueled to cook the daily meal, and mixes with all the spicy complexities of the teeming subcontinent. Movement is perpetual: cooking and haircutting and discreet bathing at public water faucets; children who respond to cameras with bright, pleased smiles; beggars murmuring incantations perhaps memorized for westerners: "Oh Mommie," a young woman carrying a baby high on her shoulder says, "oh baksheesh, oh baby, no papa, oh Mommie. . ."

India has too many people, too many poor. Yet it is beautiful and gentle and giving, and throughout the whole of the country, religion permeates everyday life without apology. "Can you wait here, please, for a small moment," the Hindu tour guide asks in his lyric, embroidered English, "while I go into the temple to have a glimpse of the Lord Shiva?"

Mumbai is said to have the largest collection of Victorian buildings in the world, many of them in a state of decay and all of them reminders of the pomp and arrogance that were part of British rule. At the university, the Rajabai Tower stands tall and elegant. Across the city, at the massive, splendid railway station called the Victoria Terminus, the dabbawallahs invade each morning, six days a week, precisely at eleven-thirty. Dabba means lunch box—the British called them tiffin boxes—the round, aluminum, tiered sort with a handle on top. Wallahs are lunch-box carriers. There are upwards of three-thousand dabbawallahs in Bombay. Their daily mission is to deliver lunch from an office worker's homes to their places of business, wherever that may be in the sprawling city, and then to return the empty boxes home again. Often this means traveling long miles by rail and roadway.

This complex operation requires the lunch to be picked up by one dabbawallah, delivered by another, and in between to be handed off by two or three others, hopping on and off trains along the way. Timing is everything; the whole system depends on the tightest of schedules, from pickup to delivery. The scene at Victoria Terminus is utter chaos, with streams of dabbawallahs wading through the crowds, carrying the boxes on their heads on narrow, two-feet-by-five-feet wooden crates. The wiry men climb off the trains and, in the frantic din, the lunch boxes are separated according to an intricate set of symbols (many dabbawallahs cannot read or write), sorted, and sent on their way—often being pushed in carts that carry as many as one hundred dabbas through the already congested city streets. The dabbawallah wears an expression of intense concentration. "Lavkar! Lavkar!" he shouts to anyone who impedes his schedule—Quickly! Quickly! His is a race against time, repeated day after day.

Miraculously, every lunch reaches its owner before one o'clock, and by two-thirty, the empty dabbas are back at the railway station on their clamorous return journey. It is a system that has been in place for more than one hundred years; in 1885, the Bombay Tiffin-Box Suppliers Association was established. The daily dabbawallah miracle is only one of the spectacles that proves the extraordinary vitality of daily life in this exotic part of the world. It is the kind of scene that gives the true traveler—one who has come to learn—pause, which forces a reconsideration of any fixed notion and blasts stereotypes about a land and its people.

Which is, in the end, what voyaging is all about. ⚓

A moment of silent reflection in Jerusalem.

Visitors to Jerusalem share the walkways with the prayer and bustle of reverent locals.

The Church of the Holy Sepulcher marks the spot where Jesus was crucified.

THE SANCTITY OF

ASHDOD

Ashdod isn't so much a destination in its own right but a gateway to the fabulous attractions of Israel. Within a short drive of the busy Mediterranean port are the ancient walled city of Akko, the undulating rural landscapes of Galilee, the sprawling Roman ruins at Caesarea, and the modern attractions of Tel Aviv. Farther afield, but still within easy reach of the coast, are such world-renowned sights as the Dead Sea, the Negev Desert, and the Jordan River. Nearly everyone who disembarks at Ashdod also makes the run up to Jerusalem, tabernacle of three great religions and one of the world's most extraordinary cities. ⚓

Port appeal rating

Adventure	★★★★
Entertainment	★★★½
Romance	★★★
Cuisine	★★★
Shopping	★★★½

Previous pages: **D***evout Hindus bathe in the sacred waters of the Ganges at Varanasi.*

T*he Taj Mahal, built by the Mogul emperor Shah Jahan to commemorate his wife, is an enduring Indian monument to love.*

B*ombay's crowded streets reflect the human mélange of modern India.*

THE FACETS OF

BOMBAY

India's largest city is the fulcrum of a national film industry that produces more films each year than any other country. Much like its California version, "Bollywood" is not so much a place as a state of mind—the glitz and glamour that envelop the movie stars who make their homes in Bombay. There are no studio tours, but you can catch a glimpse of the local product at one of the city's many air-conditioned cinemas. Big, brash masala musicals are especially enchanting. It doesn't really matter that you can't understand the Hindi dialogue —the music, passion, and action speak for themselves. ⚓

Port appeal rating

Adventure	★★★★
Entertainment	★★★
Romance	★★★
Cuisine	★★★½
Shopping	★★★★

BY GEORGIA I. HESSE

MEDITERRANEAN

SHIMMERING SEAS AND PERFUMED BREEZES

FROM VENICE, THAT GORGEOUS if shabby stage rising from a shimmering sea; from Genoa or Rhodes, the traveler sets sail today and steps ashore in yesterday: a Mediterranean cruise is a time machine.

Some five thousand years ago, the Greeks gave birth to "western" civilization on the rims of the nearly landlocked lake they called mare nostrum, our sea. She is the Helen of waters. Desired by all who eye her, she has sung a siren song to sailors from Odysseus to Onassis.

The Med is a cocktail of cultures; for cruising purposes, she encompasses several other seas: the Black, the Aegean, the Ionian, the Adriatic, the Tyrrhenian, the Ligurian, and the Alboran. Ships swim into the past from Crete in the east to Spain and Morocco in the west where, beyond the Rock of Gibraltar, they dive off the edge of the earth.

The isles of Greece! the isles of Greece
Where burning Sappho loved and sung, . . .
Eternal summer gilds them yet
But all, except their sun, is set.

BARCELONA
CIVITAVECCHIA
CRETE
GENOA
LIVORNO
MARSEILLE
MYKONOS
NAPLES
NICE
PALMA
PIRAEUS
PORTOFINO
RHODES
SANTORINI
VENICE

Lord Byron, while weeping over the decline of the classic world, walked with pleasure in the antique streets of Greece's classic cities. So wanderers do today. Perhaps the most seductive site in the Greek islands, Delphi rises from a slope of Mount Parnassus in central Greece, a short bus ride uphill above the tiny ports of Galaxidi and Itea. In the valley below waves an ocean of green-gray olive trees half as old as time. Myth meets history in the Sanctuary of Apollo where Pythia sat on a three-legged stool above a fissure in the earth and pronounced while in her trance the ambiguous utterances known as oracles.

Farther up the hill, the great theater (second century B.C.E.) and the stadium doze in the afternoon sun and bees buzz in the golden grasses. Such a sight is enough to mold anyone into a mystic.

The melancholy of brooding in ruins is nowhere more delicious than on Crete, cradle of the Minoan culture. Was the top god Zeus born in this grand,

The wisdom of the ages shows on the faces of many Greeks.

savage landscape (as Homer and Hesiod report)? Did he here love the Phoenician princess Europa? Did their son Minos construct here the first maze, the Labyrinth (Hall of the Labrys or Double Ax), and in it imprison the monstrous Minotaur? Exploring the Palace of Knossos, not far from the port of Heraklion, one comes to believe all these tales, as tightly interwoven as the threads of a tapestry.

Occasionally, the cruise passenger finds time to hike through the Gorges of Samaria, a wild canyon inhabited by ghosts and startled-looking goats.

In Mykonos, the tourists walk 'twixt souvenir shops, block by block. Mykonos, afloat in a ring of Aegean isles (the Cyclades), may be the most popular Greek island of them all, luring the jet set, the yachting crowd, the amateur painters, the nudists, the shoppers, a few leftover beatniks, and tyro photographers in search of the current incarnation of Peter the Pelican. A keen eye may spot a genuine Greek fisherman or an aproned grandmother seated on a white stoop spinning her yarn. The island remains irresistibly picturesque with its beflowered windowsills, its blindingly white alleyways, and a postcard parade of four thatched windmills. The very best thing to do on Mykonos is to settle into a sidewalk café (preferably the one facing the windmills) to consume avgolemono (a chicken-lemon soup), xofoas (grilled, oregano-flavored swordfish with green pepper), and that old favorite (it tastes new here), moussaka.

The Parthenon was dedicated to the worship of Athena, virgin goddess of ancient Greece.

Nearby Delos is a holy island, rocky and wind-beaten, where Apollo was born. Sometimes, when the fierce meltemi blows, ships can't send their tenders ashore to this Cycladian Pompeii. Of the eight or more beasts carved of Naxos marble in the seventh century B.C.E., only five still stand on the Terrace of the Lions, two of those with open mouths, roaring down the centuries. Mosaic floors, a theater, and an airy museum remain; there's no question the place is haunted.

"In Rhodes," wrote Lawrence Durrell, "the days drop as softly as fruit from trees." It is so. The Island of Roses wafts a welcome of mild weather and perfumed breezes. Grandly stands the fourteenth-century Palace of the Grand Masters, inheritor of the fortresses founded in the eleventh century by house chevaliers who protected pilgrims on their way to the Holy Land. Here and on the Rue des Chevaliers, medieval Europe remains visible—especially as enhanced by nightly son et lumière performances.

Greeks are nothing if not gregarious. When night falls on Rhodes, they throng in streets and squares to celebrate—what? The joys of being Greek, no doubt; joys they share (loudly) with every likable human being who crosses their paths. Signs of the times posted in bars, cafés, and nightclubs speak international English: "Thursday. Ladies night," they shout in red letters, "so buy two get one free." Or "If you don't have wine, how will we get to know each other?" How, indeed?

One strange day in 1470 B.C.E., about the time the waters of the Red Sea parted to let the children of Israel pass, the circular volcanic island we now call Santorini, home to a high Minoan culture imported from Crete, blew its top. At that cataclysmic moment, all life came—bang—to an end.

Today, cruise ships ease into the vast bay of crescent-shaped Santorini (also known as Thera or Thíra). In reality, they are sailing into the volcano's collapsed crater, the largest caldera on earth (twenty-eight square miles). Crescent-shaped cliffs clearly composed of volcanic layers soar for a thousand feet. Strung across the top like a white ribbon,

In the quiet of dawn, Mykonos hardly seems the jet-set haven it becomes at dusk.

THE AURA OF

MYKONOS

Despite legions of tourists over the past forty years, including many of the world's rich and famous, Mykonos has somehow managed to retain its charm. Maybe it's the whitewashed houses, the picture-postcard windmills, or the celebrated nude beaches. Or perhaps something metaphysical? That's what attracted the ancient Greeks. Nearby Delos was the hub of Greek religion, one of the spiritual centers of the Mediterranean world, a place that attracted pilgrims and apostles from far and wide—the Sante Fe of its day. ⚓

Port appeal rating

Adventure	★★★★
Entertainment	★★★
Romance	★★★★½
Cuisine	★★★½
Shopping	★★★½

the houses of the sky-city of Fira (or Thira) crowd near the rim, threatening to tumble off the view.

Traditionally, passengers reached the cramped, cobbled streets of the town by riding donkeyback up 587 zigzagging stone steps. Today, alas, a cable car makes the ascent easy. Some sailing schedules allow only time enough to take lunch or visit the handicraft shops. Others permit a tour to Akrotiri where digs of the pumice-preserved Bronze Age city began in 1967. Frescoes of a fluid grace rarely known in antiquity have been unearthed and are displayed in the National Archaeological Museum in Athens.

Is Santorini (named for its patron, Saint Irene), as believed by some scholars, really the lost island of Atlantis? Why not?

If children could create cities, they would first put one in the clouds. The second would be Venice.

Paul Theroux in 1975 likened Venice to "a drawing room in a gas station." Jan Morris in 1997 wrote that the swarming crowds "are

enough to squeeze obscenities out of a saint." But the swirl in the streets, the throngs that threaten to throw the unwary off one of the four hundred bridges, the din of tour guides in never-quiet churches, the unsightly (sometimes smelly) clutter in the canals: All this is nothing new.

For nearly eight hundred years the Serene Republic commanded an empire of world commerce, presiding over a trading cavalcade of nabobs from the nations of the world, as well as strolling strumpets, miserly mountebanks, knaves and pickpockets, and lawyers. In the

Medieval stone walls still enclose the old city of Rhodes, site of myriad shops, workshops, and museums.

THE WONDERS OF

RHODES

Rhodes is the easternmost Greek isle and in many respects the most distinct. This is largely a function of geography, its distance from Athens. But Rhodes also sports an unconventional history. For many centuries it was an independent entity, ruled by the Knights of St. John rather than the Ottoman sultans, a curious twist of fate that gives the island a cosmopolitan demeanor, more a feel of Western Europe than anything else in the Aegean. The giant bronze statue called the Colossus of Rhodes, one of the seven wonders of the ancient world, once towered above the harbor entrance. ⚓

Port appeal rating

Adventure	★★★★
Entertainment	★★★
Romance	★★★★
Cuisine	★★★½
Shopping	★★★

THE TEMPTATIONS OF

PIRAEUS

From the port city of Piraeus, it's an easy twenty-minute trip to the Greek capital. Athens is a wonderfully complex city, tempting visitors with its ancient monuments, shopping bargains, aromatic foods, and lively nightlife. Atop the Acropolis stands the ruins of the Parthenon, one of the Seven Ancient Wonders, and though hordes of visitors constantly besiege the site, you can't help but stand in awe before this great structure. At the foot of the hill lies the Plaka, a delightful warren of pedestrian streets where you can sip a glass of ouzo at a sidewalk café, search for traditional crafts in dozens of small shops, and enjoy traditional bouzouki music at a convivial nightspot. ⚓

Port appeal rating

Adventure	★★★★
Entertainment	★★★½
Romance	★★★½
Cuisine	★★★½
Shopping	★★★½

Greece's incomparable classical art and sculpture have drawn travelers to her shores for centuries.

eighteenth century, historian Edward Gibbon could write: "Old and in general ill-built houses, ruined pictures, and stinking ditches . . . a large square decorated with the worst Architecture I ever yet saw. . . ." Nonetheless, the city must be visited, her priceless paintings studied, the mosaics of St. Mark's Basilica admired, the Doge's Palace explored, the many museums prowled through, simply because Earth does not afford another Venice. (Besides, one of Europe's grandest hotels, the Cipriani, sits just across the canal on its own island.)

Naples, Civitavecchia (for Rome), Livorno (which the British call Leghorn), Portofino, Genoa: These and other seductive ports stud the boot of Italy from its toe near Sicily to the French Riviera.

"See Naples and die," says the Italian proverb, meaning one may as well perish because the globe has nothing to show more fair. Naples, however, does not reward casual callers with only a handful of hours at their disposal; it is too crabbed, too cranky, too complex. Some ships stop at Capri instead, an almost excessively pretty place that seems the setting for an operetta. From Naples, however, excursions spin off to dramatic destinations: Pompeii, the peerless ruin; Herculaneum, another victim of Mount Vesuvius's eruption in A.D. 79; the compelling Amalfi Coast, where pastel villages right out of paintings climb cliffs rising from the deep blue sea.

Amalfi, Italy's oldest republic, gave the world the Tavole Amalfitane, the original maritime code. Positano, a Moorish-looking jumble of cubic houses, is white as a cloud. On Sorrento's peninsula, a twenty-one mile road wiggles like an eel around hillsides overgrown with olive, orange, and lemon groves.

In the other direction (southeast) from Amalfi, the remains of the Greek colony of Paestum (600 B.C.E.) await: temples of a fine yellow limestone, ruins of dwellings shaded by cypresses, the superb Doric Temple of Neptune, a Roman amphitheater, and a museum housing handsome Greek funerary paintings.

Guards in traditional "tsolias" uniforms march at Sintagma Square in Athens.

Travelers reach Ravello, a treasure of a town perched between sea and sky, by a road that hairpins up the narrow Dragon Valley from the fishing village of Atrani. Its two historic villas welcome visitors (but Gore Vidal's does not); the view, especially from the Hotel Palumbo, remains the pièce de résistance of the whole coast.

Go thou to Rome—at once the Paradise, The grave, the city, and the wilderness. To visit this capital, celebrated by Shelley, one might take an excursion bus, train, or taxi from Civitavecchia, the port for Rome since the reign of Trajan. However, a mere few hours spent with its antiquities may breed only frustration. Better to hire a car to drive to the small town of Tarquinia and/or Cerveteri, each a fascinating Etruscan necropolis.

More than two thousand Greek islands await exploration by the adventurous.

Livorno serves as a steppingstone for the two splendid cities of Pisa and Lucca. The fame of Pisa falls at the feet of its leaning tower, the white marble Romanesque campanile, although the striped light-and-dark marble cathedral and the majestic baptistery are equally admirable.

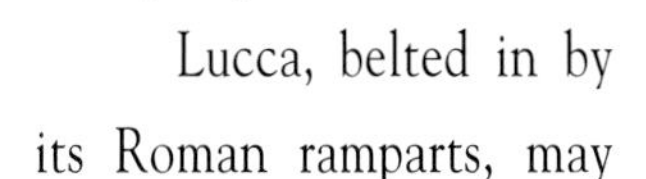

Lucca, belted in by its Roman ramparts, may be the city that first-time visitors to Italy have dreamed of. Although less rich in works of artistic genius than Florence, Lucca's Old Town remains so far unpolluted by the plagues of mass tourism and retains a purity of Pisan architecture.

Northwest, along the Riviera di Levante, Portofino snoozes in the sun, backed by a rugged promontory. "Hang out here, around my

Umbrellas undulate across a Rhodes beach.

THE MYSTERIES OF

SANTORINI

With whitewashed houses set against stark volcanic cliffs, Santorini is the most handsome island in the entire Aegean. It's also one of the most mysterious. Minoan settlers came here from Crete around 1500 B.C.E. and created a splendid city called Akrotiri. Most of the population perished when Akrotiri sank into the sea during a massive volcanic eruption that tore the island apart. This calamity probably gave rise to the legend of Atlantis. But at the same time, volcanic forces created the stunning landscape that makes Santorini such a special place today. ⚓

Port appeal rating

Adventure	★★★★
Entertainment	★★★
Romance	★★★★½
Cuisine	★★★½
Shopping	★★★½

THE APPEAL OF

CRETE

Crete's Minoan heritage consumes the interest of most visitors to this largest of Greek islands. Whether they dock in Heraklion or Ayios Nikolaos, visitors' prime targets are the colorful reconstructed Palace of Knossos, where King Minos confined the legendary Minotaur, and the Minoan treasures in Heraklion's Archaeological Museum. More adventuresome visitors head for the spectacular Samaria Gorge, the longest in Europe. Starting with a precipitous trail that drops 3,280 feet in the first mile, hikers pass by tiny chapels, abandoned villages and squeeze through the Iron Gates (towering rock walls only nine feet apart) on their eleven-mile walk to the sea. ⚓

Port appeal rating

Adventure	★★★★
Entertainment	★★★
Romance	★★★★
Cuisine	★★★½
Shopping	★★★

The Palace of Knossos on Crete recalls the glory of the ancient Minoan civilization.

Seabourn Legend *passes below Santorini's main town, perched on the edge of a volcanic escarpment.*

The Ponte Vecchio leaps the River Arno at Florence.

harbor," it seems to say. "Sip a little something." So small (population 611), so pretty, so undemanding is it that many passengers sit right down at a café and stay until the ship sails.

Genoa is Italy's finest seaport, contained in a natural, mountainous amphitheater. Often scorned for some reason by American travelers, it owns one of the loveliest streets in Italy, the Via Garibaldi in the Old Town, lined by regal sixteenth-century palaces, many of them home to art galleries and museums.

In the late 1960s, when Brigitte Bardot removed her itsy-bitsy bikini and turned her glutei maximi to the smiling sun of St. Tropez, the French Riviera began another incarnation as "the sunny place for shady people." The Riviera (including the principality of Monaco), "invented" in 1834 by Henry Peter, England's Lord Brougham, still reigns as the greatest, fizziest seaside resort; a place the world comes to lark about in the sun and to live la dolce vita.

The daze of wine and roses lives on in Monaco, say, or Villefranche or Nice. It is impossible not to become intoxicated by the look of olive groves, of pine and oak forests scaling slopes crowned by antique villages with twists for streets; by tastes of garlic and oil and fresh fish and pink lamb and Rhône wines; by scents of lavender and tea roses and jasmine and thyme and rosemary and sage. Where nature herself is a painting, artwork feels at home: Thirty-some outstanding museums, artists' houses and galleries, decorated chapels, and more delight the eye between Menton and St. Tropez.

Florence's Duomo cathedral is a crowning achievement of Renaissance design.

Twenty-six centuries ago, Marseille was planted on the French coast by Greeks from Asia Minor. It always has been France's greatest port, but in the decades following World War II it decayed; nice people didn't want to wander there. (A few gadabout gourmets still arrived in search of a smashing bouillabaisse.) Today, smart cruise liners call on a spiffy, spruced-up city that looks forward to the completion of Euroméditerranée, a dynamic economic project intended to re-create the glory days of the mid-nineteenth century.

For cruisers, the Mediterranean begins or ends in Barcelona: exhaustingly energetic, infectiously gleeful Barcelona, as eternally fanciful as a Picasso portrait. Along the Ramblas, street life plays itself out until early morning (nobody dines in Barcelona until eleven P.M.). Where Ferdinand and Isabella received Columbus on his return from the New World, rock musicians costumed as motorcycle thugs shatter the silence of the small Plaza of the King.

The Mediterranean, mare nostrum: old as Neptune, new as dawn. The ship's whistle bellows a haunting farewell.

There's not a dry eye on the deck. ⚓

Neptune's Fountain dominates the Piazza della Signoria in Florence.

Only a few hundred gondolas remain from a Venetian fleet that numbered more than twenty thousand.

THE LURE OF

VENICE

"Nothing in the world that you have heard of Venice is equal to the magnificent and stupendous reality." So wrote Charles Dickens in 1844, one of numerous artists, writers, and rascals who have been lured to the Queen of the Adriatic. Casanova frolicked here with a number of local ladies. Mark Twain dubbed it "a funny old city" where legs were obsolete —because gondolas could take you anywhere. Lord Byron disregarded the boats, preferring to swim the Grand Canal from end to end. Thomas Mann found it a rather melancholy place, the perfect inspiration for his *Death in Venice*. Ralph Waldo Emerson poked fun, declaring that Venice was "a city for beavers." ⚓

P*ort appeal rating*

Adventure	★★★★½
Entertainment	★★★★
Romance	★★★★★
Cuisine	★★★★½
Shopping	★★★★

THE CHARM OF

PORTOFINO

It's easy to while away an entire day lounging in the waterfront cafés of Portofino. But this is also an active port of call. The picturesque Portofino Peninsula is ideal for walking. Trails lead along the rugged coast or through pine and olive groves to small villages. One of the easiest walks is the one-hour jaunt to the lighthouse overlooking the Gulf of Rapallo. The tiny fishing village of San Fruttuoso with its Benedictine abbey is roughly four hours away by foot. ⚓

Port appeal rating

Adventure	★★★★
Entertainment	★★★
Romance	★★★★½
Cuisine	★★★★½
Shopping	★★★

Picture-perfect Portofino shimmers in the summer Mediterranean sun.

THE VISTAS OF

NAPLES

As one of Europe's last active volcanoes, Mount Vesuvius is a constant source of wonder and even fear. Its swarthy cone rises to a height of thirty-seven hundred feet above the Bay of Naples, an ever-present reminder that Mother Nature remains untamed. A paved road leads to a parking lot on the mountain's northern flank; from there you can walk or ride a chairlift to the summit. The views are magnificent—a sweeping panorama of Naples, Capri, Ischia, and the Sorrento Peninsula. If you long for a close-up look at volcanic destruction, you can visit the ruins at Pompeii or Herculaneum, once-great Roman cities destroyed by the eruption of A.D. 79. ⚓

Port appeal rating

Adventure	★★★★
Entertainment	★★★
Romance	★★★½
Cuisine	★★★★
Shopping	★★★

THE INTRIGUE OF

PALMA

Unlike much of Mallorca with its flashy hotels, the city of Palma retains a good deal of its bygone ambience, especially the dark and brooding Barrio Gotico. This maze of narrow cobblestone streets is ideal for walking and ripe for exploration. Moorish influence is manifest in structures like the Almudaina Palace and the Arab Baths, while the towering cathedral boasts a Gothic facade and Romanesque cloister. Reflecting the Balearic Islands's close ties with Catalonia, the cathedral bears a huge wrought-iron sculpture by Gaudí rather than a traditional altar. ⚓

Port appeal rating

Adventure	★★★½
Entertainment	★★★
Romance	★★★★
Cuisine	★★★
Shopping	★★★

Palma's Belver Castle maintains its bygone splendor.

Left: The goddess Diana surveys the remains of Pompeii.

The art of Pompeii's art of Casa del Fauno still reflects the glory of ancient Rome.

The Trevi Fountain has long attracted lovers and art lovers from around the world.

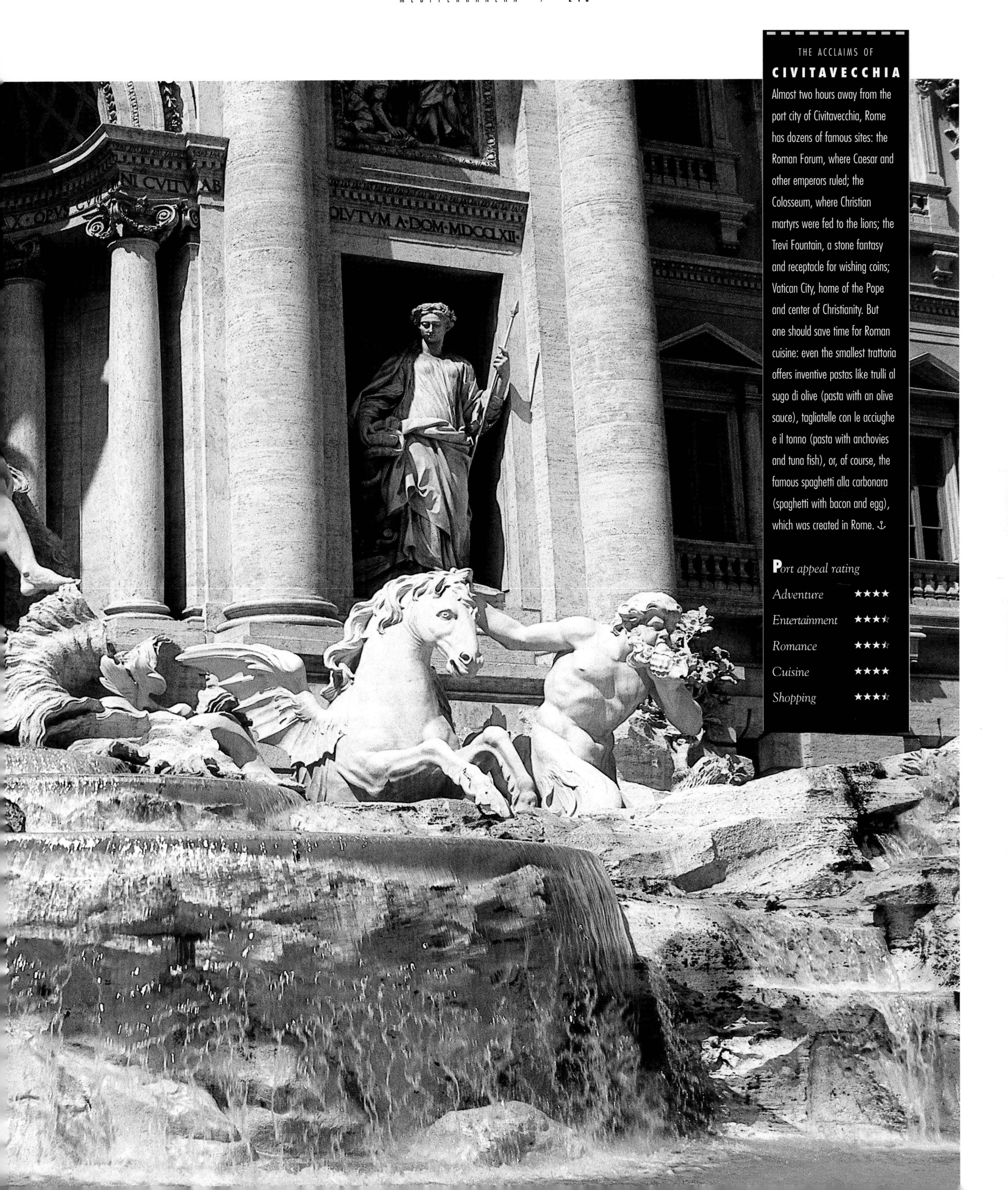

THE ACCLAIMS OF

CIVITAVECCHIA

Almost two hours away from the port city of Civitavecchia, Rome has dozens of famous sites: the Roman Forum, where Caesar and other emperors ruled; the Colosseum, where Christian martyrs were fed to the lions; the Trevi Fountain, a stone fantasy and receptacle for wishing coins; Vatican City, home of the Pope and center of Christianity. But one should save time for Roman cuisine: even the smallest trattoria offers inventive pastas like trulli al sugo di olive (pasta with an olive sauce), tagliatelle con le acciughe e il tonno (pasta with anchovies and tuna fish), or, of course, the famous spaghetti alla carbonara (spaghetti with bacon and egg), which was created in Rome. ⚓

Port appeal rating

Adventure	★★★★
Entertainment	★★★½
Romance	★★★½
Cuisine	★★★★
Shopping	★★★½

THE MARVELS OF

LIVORNO

Livorno (or Leghorn) is the chief port of Tuscany, that marvelous Italian region known for fine art, exquisite food, and divine vistas. As such, the city is a gateway to various other cities in the Arno Valley, including tourist meccas like Pisa, Siena, and Florence. The Medici rulers of Florence—trying to compete with their archrivals in Venice—developed Livorno into a major port starting in the late sixteenth century. The Medici funded construction of a stone jetty, protective bastions, and a barge canal to Pisa. Livorno's other claim to fame is the Italian naval academy. ⚓

P*ort appeal rating*

Adventure	★★★★½
Entertainment	★★★½
Romance	★★★★
Cuisine	★★★★½
Shopping	★★★★

THE MASTERS OF

GENOA

If you have time to explore only one place in Genoa, make it the Via Garibaldi in the heart of the old town. Often called the street of palaces, Garibaldi is flanked by fine Renaissance structures built during Genoa's golden age. Palazzo Doria Tursi, the city hall, features displays on famous Genoese, including a violin that once belonged to Paganini. Palazzo Bianco and Palazzo Rosso harbor the city's best art museums, especially strong on local painters and Flemish masters like Rubens and Van Dyck. Palazzo Cataldi is famous for its Renaissance murals and golden hall. And the Piazza Fontane Marose is the perfect place to rest your weary feet. ⚓

Port appeal rating

Adventure	★★★½
Entertainment	★★★
Romance	★★★
Cuisine	★★★★
Shopping	★★★

Genoa's Old Town is a maze of medieval streets and plazas.

Among the artistic treasures of Florence is the remarkable Uffizi Gallery.

Overleaf: Luxury yachts and Seabourn Legend jockey for berths along the Monte Carlo waterfront.

THE STYLES OF

NICE

The French Riviera offers more than a hundred different beaches; the trick is choosing one to fit your mood on any given day. The best for people watching are La Croisette in Cannes (especially during the film festival) and the rocky shore along the Promenade des Anglais in Nice. For fine white sand (and bronzed bodies) try the Plage des Salins in St. Tropez. Some of the most picturesque strands are the least crowded, including Saint-Raphaël, Beaulieu, and Cap d'Ail. Mougins, where Picasso spent his final years, doesn't have much in the way of sand. But the coast here is simply irresistible, the sort of beauty that kindles the artist in all of us. ⚓

Port appeal rating

Adventure	★★★★½
Entertainment	★★★★
Romance	★★★★½
Cuisine	★★★★½
Shopping	★★★½

THE NUANCES OF

MARSEILLE

Cruise ships dock at modern La Joliette, but the heart and soul of Marseille is still the Vieux Port (Old Port), where fishing boats jostle for space with million-dollar yachts. A lively fish market still unfolds each morning along the Quai des Belges. More than three thousand years of local history cluster nearby: Greek and Roman ruins in the Jardin des Vestiges, the medieval St. Ferreol Church, and the Marseille History Museum. It's perfectly acceptable to slurp the best bouillabaisse in France in the restaurants along the Quai de Port. Or you can take the funky little ferry to the opposite shore where you can explore the trendy waterfront cafés and shops of the Cours d'Estienne d'Orves redevelopment zone. ⚓

Port appeal rating

Adventure	★★★★
Entertainment	★★★
Romance	★★★
Cuisine	★★★★
Shopping	★★★

The Vieux Port of Marseille is the entryway to the many charms of Provence.

Art nouveau master Antonio Gaudí left his indelible mark on Barcelona's Sagrada Familia Church.

THE INSPIRATIONS OF

BARCELONA

Barcelona's hinterland—the Spanish region of Catalonia—has given the world some of its most renowned modern artists. Pablo Picasso, Salvador Dalí, and Joan Miró were all native sons, as was the inimitable Antonio Gaudí. The others moved on to fame and fortune abroad; but Gaudí remained in his beloved Barcelona, where most of his most outstanding works can be seen. The marvelous Sagrada Familia Church, still unfinished more than a century after its cornerstone was laid, is a masterpiece of art nouveau. The Paseo de Gracia and Parc Güell are more sublime statements of his singular style. ⚓

Port appeal rating

Adventure	★★★★½
Entertainment	★★★★
Romance	★★★★
Cuisine	★★★★½
Shopping	★★★★

BY KATE SEKULES

E·U·R·O·P·E

A CONTINENT OF CONTRASTS IN CULTURE AND TIME

AMSTERDAM
BERGEN
BORDEAUX
COPENHAGEN
DUBLIN
DUBROVNIK
HAMBURG
LISBON
OSLO
STOCKHOLM
ST. PETERSBURG
TILBURY
TURKU

THERE IS NO GENERALIZATION to be made about European ports. Venice, la Serenissima, the maritime city-state that once dominated the entire world and still dominates romantic fantasies, bears no relation whatsoever to, say, Bergen, Norway's second-largest city. Bergen, which the European Union (EU) has named the Center of European Culture in 2000, has the same size population as Baton Rouge; however, Bergen rests its elbow not on the bayou, but on the breathtaking million-year-old fjords of Norway's west coast. No, Bergen has nothing in common with Venice, nor with Tilbury, the nondescript dock for London, nor with cosmopolitan Hamburg, the Hanseatic port. . . .

Hold it. Bergen and Hamburg were indeed once connected, since Bergen was also a member of the medieval Hanseatic League. Hamburg, in turn, has its Venetian parallel. Hamburgers describe their city as the German Venice, since it not only teeters on the water's edge, but invites some of it to flow right on through: the great river Elbe and the smaller Alster with its dammed lakes, Binnenalster and Aussenalster, snake through this seaport city.

The point is that European ports, like Hamburg, Bergen, and Tilbury, introduce and represent not only regions, but discrete cultures, each with its own language, its centuries of history, art, architecture, cuisine, and way of life. Each country's spirit is encoded not only in its famous buildings, but in its quotidian details—the Spanish never sup before ten, for instance, while their neighbors, the Portuguese, dine at eight; the French still consider a three-hour, three-course lunch their birthright, while the British nowadays grab a sandwich at their desk.

A European voyage means sailing from the midnight sun or frigid, black winter of the far north to sunbaked Lisbon; it means ordering smörgåsbord, paella, bouillabaisse, rijsttafel, caldeirada, or fish and chips, washed down with a glass of akvavit, Chateau Margaux, slivovitz, or Guiness. A ship avoids the "if it's Tuesday, this must be Belgium" pitfall of faster

Norway's National Day (May 17) brings out the best of local traditions and costumes.

transportation and offers fusion with the history encoded in these cities and cultures. Its port, after all, was each country's opportunity for growth and change, the cross-fertilization point where foreign ideas entered and goods were exchanged, and also the vulnerable mouth that had to be closed to invasion.

Across the continent, these natural bays, fortified, became the focus of power and occasionally remained that way. Although defense now has nothing to do with ports, it is still thrilling to dock in one of the capital cities of Europe, especially when the remnants of a former heyday are visible. Portugal's capital, Lisbon, is one of these palimpsests, where era quite obviously overlays era. We forget this—if we ever knew—but the Portuguese were the chief explorers of the world, the rich and famous of the fifteenth century, especially when Henry the Navigator was at the helm during the time they call the *descobrimentos*. Perhaps loss of glory accounts for the songs of pain and bleeding heart, the fado, the world's most agonizing musical form, performed—not only for tourists—nightly in the wine shops of Lisbon's medieval old town, the Alfama. These tangled alleys of white-walled, red-roofed houses, along with the Romanesque cathedral and the most imposing landmark you can see from the ship, the Moorish Castelo de São Jorge, were all that survived the calamitous 1755 earthquake.

A mosaic in the Lisbon sidewalk recalls Portugal's seafaring past.

On top of the Alfama, (chronologically speaking, but geographically below it) is the Baixa, or Lower Town, a lovely feat of neoclassical town-planning that predates Haussmann's more famous Parisian design. Here, wide boulevards are lined with black-and-white calçada, mosaiclike cobblestones, and houses are faced with azulejos, the blue-on-white ceramic panels reminiscent of delft pottery. One of the best things to buy in Lisbon is pottery, handpainted like azulejos or multicolored and floribundant, like the balconies of the Alfama. And the best things to eat are served on that pottery—unpretentious dishes like the ubiquitous seafood stew, caldeirada; or the national dish, bacalhau, salt cod prepared, they say, 365 ways; or simply sardinhas assadas —grilled fresh sardines. Shopping is best in the Baixa; eating in the Bairro Alto, or Upper Town, reached by one of the three funicular "elevators," the Elevador da Glória. The most extraordinary of the three, however, is the Elevador de Santa Justa, a Gothic iron tower that brings to mind the best-known of all European landmarks, the Eiffel Tower—no doubt because it was designed by Raul Mesnier, a Portuguese pupil of M. Gustave Eiffel himself.

The peacefulness and the cool, natural beauty along Norway's incredible fjords are unlike any other part of the world.

That is one joy of cruising Europe: to spot congruence where it is least expected—an Eiffel-like tower and Haussmann-like district in a most un-Parisian city. Another joy is the opposite of that, to dock in a world where everything stands in starkest contrast to the foregoing—as it does from Lisbon to Amsterdam. (Portuguese Sephardic Jews made this voyage during the fifteenth and sixteenth centuries; you, however, will stop several times en route.) The Dutch capital is, of course, arranged around canals like Venice, but there the similarity ends. The Netherlands never courted power, but always had good economic luck and judgment, something the atmosphere of Europe's most laid-back port capital reflects.

Lounging in a bruine café—yes, brown café—over a beer; bicycling carefully over cobblestones and bridges; ordering a rijsttafel, a dish from Indonesia (a Dutch colony) of

More than sixty-seven hundred historic buildings are preserved along Amstersdam's canals.

THE AVENUES OF

AMSTERDAM

Amsterdam was born in the thirteenth century when a small dam was built at the mouth of the Amstel River to keep back the North Sea. Since then the city has expanded in concentric horseshoe-shaped wedges of water and land, an intricate maze of canals and islands that give Amsterdam its unique feel. Some of the canals have been filled in to create major streets like the Damrak and Rokin. But there is still plenty of water: the city boasts more than a thousand bridges and several thousand houseboats. One of the best ways to explore the waterways, to really get to know the city, is a canal boat tour. ⚓

Port appeal rating

Adventure	★★★★
Entertainment	★★★★
Romance	★★★½
Cuisine	★★★★
Shopping	★★★★

THE ATTRACTIONS OF

LISBON

As one of Europe's leading commercial ports for more than five-hundred years and the origin of many voyages of discovery to the New World, Lisbon is no stranger to the sea. But until quite recently little of the city's tourism revolved around the waterfront. That all changed in 1998 when the Portuguese capital hosted its first world expo. Arrayed along the banks of the river Tagus, the expo endowed Lisbon with several permanent attractions, including a virtual reality pavilion with four thematic settings, a multipurpose sports and cultural center, and Europe's largest aquarium with a massive central tank as large as four football fields. ⚓

Port appeal rating

Adventure	★★★★
Entertainment	★★★½
Romance	★★★★
Cuisine	★★★★
Shopping	★★★½

Capo de Roca is the farthest point west in Portugal.

Great wine aside, Bordeaux is also a bastion of superb French cuisine.

THE ENTICEMENTS OF

BORDEAUX

Some of the world's most celebrated wines are made in and around Bordeaux, the world's largest fine wine region, with many of them taking their names from the towns in the region: St. Emilion, Medoc, Sauterne, Pomerol, Margaux. Many chateaux are open for tastings during the summer months. More than forty-four million cases of wine are produced annually in the region. ⚓

P*ort appeal rating*

Adventure	★★★★
Entertainment	★★★
Romance	★★★½
Cuisine	★★★★½
Shopping	★★★½

THE INTRICACIES OF

TILBURY

Most seaborne visitors to the British capital, served by the port city of Tilbury, head straight for landmark sights like Big Ben and Buckingham Palace. But London's docklands boast more than their fair share of attractions. The Royal Observatory at Greenwich sits astride the prime meridian, from which all of the world's longitude and time zones are measured. The National Maritime Museum, which details more than a thousand years of British seafaring history, is housed within the sublime confines of the Queen's House, a masterpiece of Palladian architecture. The last of the great clipper ships, the majestic *Cutty Sark*, anchors along the Greenwich waterfront. ⚓

Port appeal rating

Adventure	★★★★½
Entertainment	★★★★½
Romance	★★★★
Cuisine	★★★★
Shopping	★★★★

rice surrounded by twenty spicy side dishes; viewing the Rembrandts in the Rijksmuseum, the van Goghs next door, French impressionists, and American pop and action artists in the Stedelijk Museum; listening to jazz; openly buying and smoking hashish; hailing a water taxi; scouring the flea market on the Waterlooplein—these are all activities of the Amsterdammer as much as the visitor. Meeting the locals is easy, since everyone is fluent in English, loves to practice, and suffers no attitude problem. This is especially noticeable if the next stop on the ship's itinerary is London or Paris, neither of which is noted for modest, eager-to-please natives.

To reach either great metropolis by ship requires a land journey of one to three hours—from Tilbury or Southampton for London, and from LeHavre for Paris. Both are unmissable cities but only London is maritime. British history is all about conquering the seas, though visitors to England's capital today never notice. Unless, that is, they hop on a smaller boat to Greenwich, eight miles downstream from the South Bank Center, home of the Royal National Theatre. Strung along this stretch of the Thames is a patchwork impression of what London was when Britain earned the epithet "Great" and what the city is now, as it evolves into the European center of fun and food and fashion. Here, on the left, representing the historic, the distinctive green dome of Christopher Wren's St. Paul's is followed by the Tower of London, seat of power for nine centuries, while the right-hand side—the formerly boring South Bank—is littered with signifiers of what Europe means today.

The Royal Viking Sun *sails past the wild and wind-swept Scottish coast.*

Here is Shakespeare's Globe Theater, built precisely to the design of the Bard's original "Wooden O," and the new Tate Gallery of Modern Art, a fiercely minimal interior secreted inside the industrial revolution shell of the former Bankside Power Station; then there's the OXO Tower, one of the countless hot restaurants (grilled Dover sole with sea urchin butter or roast saddle of hare with port and juniper sauce?), and the Pont de la Tour (salade niçoise or Caesar salad? Gravadlax or haddock fish cakes?) abutting the familiar Victorian-Gothic turrets and bascules of Tower Bridge. Greenwich itself has bucolic parkland, the Maritime Museum, and the Old Royal Observatory, where you can literally straddle the globe, with one foot in the right hemisphere, the other in the left, across the brass line that the earth's clocks observe as the site of Greenwich Mean Time (GMT). Most symbolic of all (so the taxpayers, who invested billions via the national lottery, hope) stands the humongous UFO that is the Millennium Dome, Europe's newest and largest landmark.

The port city of Amsterdam, a rich blend of the traditional and modern, is the first taste of Holland for many cruisers.

Britain has never been so integrated into Europe as it is today, a phenomenon that is partly explained by its membership in the EU, partly by a significant landmark of no interest to the ocean voyager, the Channel Tunnel.

Farther north still, the Scandinavian countries, EU members all, have also intermittently entertained a certain distance from their neighbors. Tempting as it is to consider Sweden, Denmark, and Norway as one big Scand nation, each country is really quite distinct, as you discover by docking in their respective capitals, Stockholm, Copenhagen, and Oslo. What do these cities have in common? An openness and optimism, especially in the light summer months when play is the imperative; social democratic politics that ensure an enviable standard of living; the clean lines of Scandinavian design, now back in vogue; and a similar cuisine, especially smörgåsbord (Sweden), smørrebrød (Denmark), and smørbrød (Norway).

Stockholm, more so than Amsterdam, Hamburg, even Venice, is, "the city that floats on water," spread as it is over fourteen islands of an archipelago of some twenty-five thousand more. Messing about

Beefeaters have guarded the Tower of London since the time of Henry VIII.

Big Ben's minute hand is as tall as a London double-decker bus.

on boats is what everyone does come summer. Stockholm has further surprises: the perfectly preserved thirteenth-century old town, Gamla Stan, nudges up to the gigantic Stadshuset, or city hall, a 1923 collaboration of National Romantic Movement artists where, in the Gulden Hall, the Swedish Academy throws its Nobel Prize banquets, dazzling the world with its nineteen million gold mosaic tiles. The subway that brings you there (all transportation and museums are free with a Stockholm Card) is also a collaboration, with half of its ninety-nine stations bemuraled like a Tolkien tale by some seventy artists. On the island of Djurgården, you can stroll the history of rural Sweden in the form of 150 reconstructed houses at Skansen. You'll want to make notes for home: after all, IKEA was born here—not to mention Orrefors glass, the Volvo, and the Saab.

Copenhagen also spreads over islands, two of them: Slotsholmen and Christianshavn. In this, Scandinavia's most affordable capital (though this is relative), the galleries of Georg Jensen silver and Royal Copenhagen

porcelain in the trendy shopping and strolling neighborhood, Strøget, are essential stops. So are the famous Tivoli Gardens, which offer fireworks three nights a week, fairy-lighted trees, and the Ny Glyptotek (new picture gallery) with its palm trees, lunchtime concerts, and fine Egyptian and Etruscan sculptures. The number and variety of Copenhagen's museums speak of its cosmopolitan soul. Here are collections of musical history, erotica, Viking artifacts, theater, geology, art (of course), and the works of Hans Christian Andersen, plus the Guinness World of Records and

What better place to meet the Irish than in one of Dublin's traditional pubs?

THE POETRY OF

DUBLIN

The Emerald Isle is known for song and dance. But its largest city is firmly entrenched in arts and letters, having produced more great authors and poets than just about any other city in the English-speaking world. Among legendary bards either born or bred in Dublin were Jonathan Swift, Oscar Wilde, George Bernard Shaw, William Butler Yeats, and the beloved James Joyce. It was Joyce who put Dublin on the world literary map in books like *Finnegan's Wake* and the incomparable *Ulysses* which charts Leopold Bloom's nineteen-hour odyssey across the city in search of both himself and the meaning of life. ⚓

Port appeal rating

Adventure	★★★★
Entertainment	★★★★
Romance	★★★½
Cuisine	★★★½
Shopping	★★★½

Ripley's Believe It or Not museums. The Andersen museum is missable; the *Little Mermaid* statue in the harbor is not. The other most famous Danes are Carlsberg, the beer (take the free brewery tour) and Hamlet, the prince, whose Helsingør Castle is a short train ride away.

At the top of the "famous Norwegians" list are Henrik Ibsen, and the Edvards Grieg and Munch—and the Munch-Muséet is, needless to say, a star sight of Oslo. Listening to Grieg tips you off to the mystical, martial nature of this ancient northern land of the Vikings, whose farthest shores never get dark in summer nor light in winter and where reindeer farming is a career option. Oslo is one of the largest capitals in the world, with city limits consisting not of suburban sprawl, but of fjords, forests, and mountains, though a mere half a million call it home. Downtown is compact, with the big square—the Studenter-lunden, everyone's gathering place—encircled by the national theater, the university, the national museums, and the parliament, Stortinget; then, toward the waterfront, by the red brick Rådhuset, or city hall, with the medieval castle, Akershus Slott, facing it.

Since a voyage around these seas dictates a great deal of activity, you will by now have become an expert at slowing the clock. The transatlantic voyage home won't force you to abandon this skill. The trick is holding on to your new-found freedom after you dock in home port. But that's what your next voyage is for. ⚓

The Little Mermaid *keeps a calm eye on cruise ship traffic in Copenhagen harbor.*

Beautifully hewn wood constructed churches and houses are a charming characteristic of southern Norway.

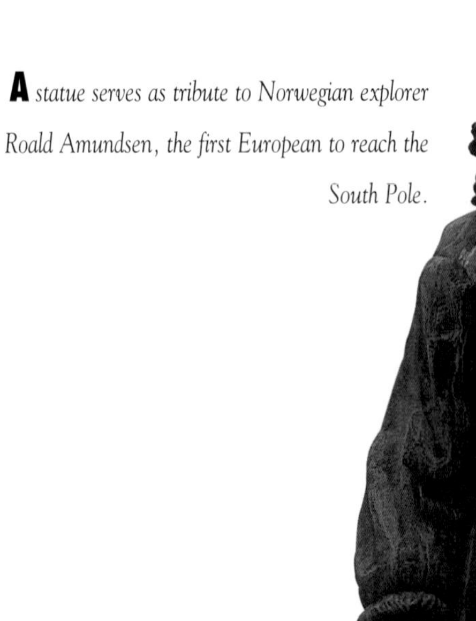

A statue serves as tribute to Norwegian explorer Roald Amundsen, the first European to reach the South Pole.

THE FESTIVITIES OF

BERGEN

Bergen likes to think of itself as the cultural capital of Scandinavia, a city in which music, dance, and drama take precedence over more tactile traditions like sauna or ski jumping. The highlight of the local cultural calendar is the Bergen Festspill, an international arts festival that spreads across two weeks in late May and early June. But there's some sort of event nearly every day, especially during the summer months when the twilight stretches well into the evening. ⚓

Port appeal rating

Adventure	★★★★
Entertainment	★★★
Romance	★★★½
Cuisine	★★★½
Shopping	★★★

Brightly painted wooden houses are one of Bergen's trademarks.

THE SPLENDOR OF

STOCKHOLM

Sweden's capital spreads across fourteen islands with their own attractions and dispositions. The city first took root on historic Gamla Stan where pastel houses crowd along narrow cobblestone streets. Skeppsholmen boasts an avant-garde landmark: the Museum of Modern Art, one of Europe's foremost showcases for twentieth-century painting and sculpture. Djurgarden is the city's premier green space with sprawling lawns, woods, and a small zoo. If you run out of urban islands, board a ferry for the nearby Skargard Archipelago, which features thousands of bits and bobs of land, some of them scarcely inhabited. ⚓

Port appeal rating

Adventure	★★★★½
Entertainment	★★★½
Romance	★★★★
Cuisine	★★★½
Shopping	★★★½

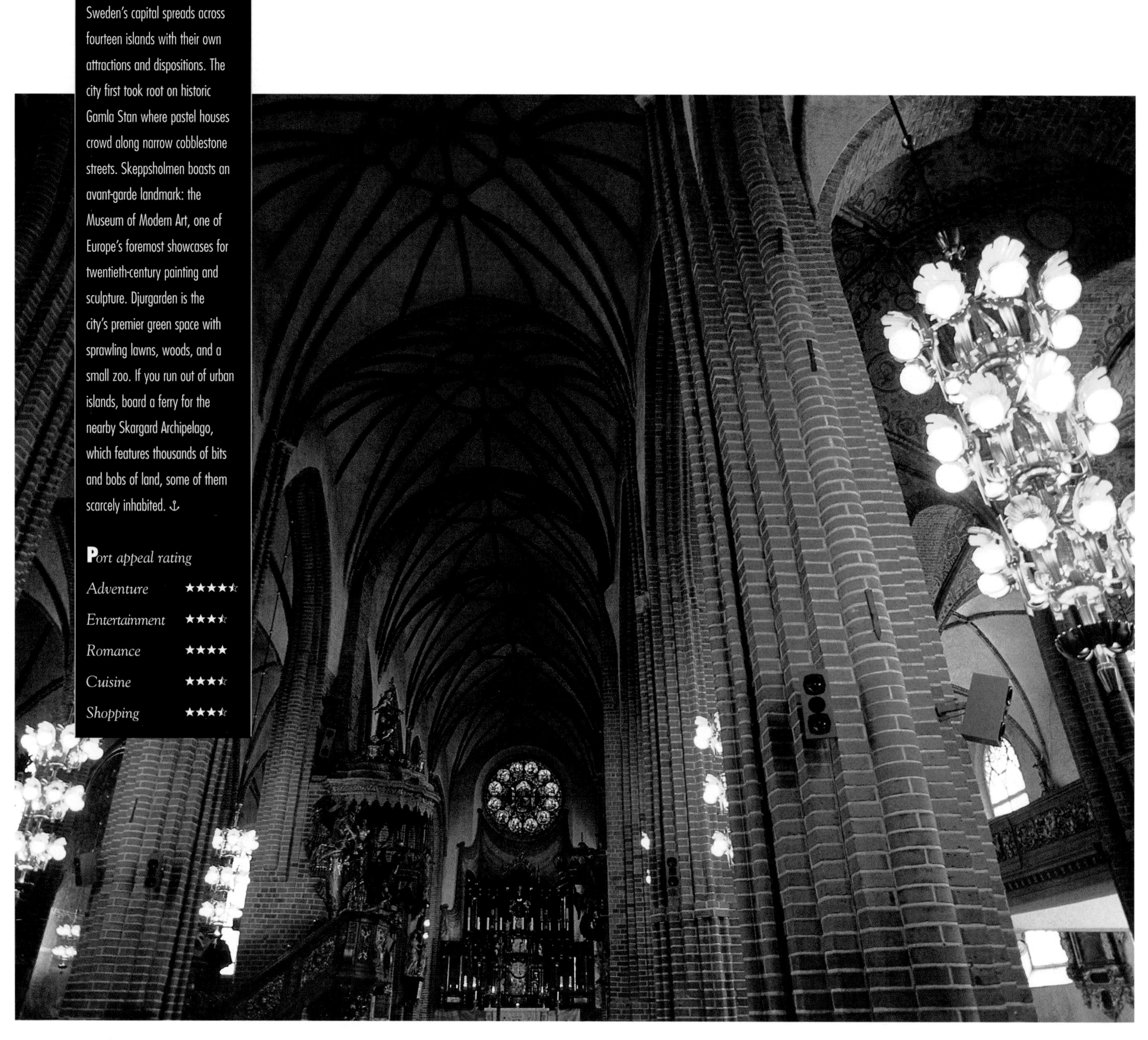

Visitors to Stockholm take in the majestic Storkyrkan where Sweden's monarchs have traditionally been crowned.

The spirit of the Vikings is kept alive in Norway's many museums.

THE FLAVORS OF

OSLO

"Everything is fishy," wrote a nineteenth-century British traveler during a visit to Norway. "You eat fish and drink fish and smell fish and breathe fish." Oslo has since cleaned up its act and is now one of Europe's tidiest cities. But seafood is still a mainstay of local menus, fresh fish that literally comes straight from Scandinavian rivers and the North Sea. A typical Norwegian breakfast always includes sild (herring) with eggs, cheese, and flatbro crackers. A favorite hors d'oeuvre is gravadlax (raw salmon), washed down with a shot of akvavit. Main courses include orret med agurken (trout with cucumbers) and rodspette (filet of plaice). ⚓

Port appeal rating

Adventure	★★★★
Entertainment	★★★
Romance	★★★
Cuisine	★★★
Shopping	★★★

Old salts and young sailors gather along Copenhagen's Nyhavn Canal.

THE TRAILS OF

COPENHAGEN

Numerous pedestrian streets and plazas make Copenhagen a paradise for walkers. The Danish capital is also easily explored by bicycle, which can be hired at reasonable cost at numerous places around the central city. One of the more popular hike/bike routes is a square-shaped circuit that takes you along the Frederiksholms Canal and three other inland waterways with stops at Christianborg Palace and several outstanding museums. The more adventurous can pedal north along the coast to Helsingør, home of Hamlet's castle and the Louisiana Museum of Modern Art. ⚓

Port appeal rating

Adventure	★★★★
Entertainment	★★★★
Romance	★★★★
Cuisine	★★★★
Shopping	★★★★

THE SPIRIT OF

TURKU

Helsinki, served by port city Turku, flaunts a split personality: a staid northern capital firmly entrenched in the past and a bold modern metropolis that has more in common with the twenty-first century. The city's sober facade is the legacy of nineteenth-century German architect Carl Engel, who transformed Senate Square and environs into a neoclassical masterpiece. The city's flamboyant demeanor is the work of twentieth-century avant-garde Finnish architects who experimented with startling combinations of wood, stone, and glass. ⚓

Port appeal rating

Adventure	★★★★
Entertainment	★★★
Romance	★★★
Cuisine	★★★½
Shopping	★★★½

Helsinki's cathedral looms over stately Senate Square.

Hamburg serves as a maritime gateway to the glories of reborn Berlin.

THE SOUNDS OF

HAMBURG

Vienna may lay claim to the title of Europe's music capital, but Hamburg runs a close second. Classical composers Johannes Brahms and Felix Mendelssohn were both Hamburg natives and today the city boasts no less than three symphony orchestras. Another bastion of culture is the city's world-renowned State Opera, largest of two dozen theaters that offer everything from Bavarian cabaret to Shakespeare. The notorious Reeperbahn is where a fledgling rock'n'roll band called the Beatles first earned plaudits. Germany's punk rock movement kicked off here, and Hamburg is now known as the country's hotbed of jazz. ⚓

Port appeal rating

Adventure	★★★½
Entertainment	★★★½
Romance	★★★
Cuisine	★★★½
Shopping	★★★

THE VISION OF

ST. PETERSBURG

St. Petersburg owes its existence to the vision of a single man: Russian czar Peter the Great, who wanted to create a "Window on the West" along the shores of the Baltic. Peter founded the city himself, in 1703, on marshy wasteland at the mouth of the Neva River. Drawing inspiration from Venice and Amsterdam, the new metropolis took shape around a network of canals and islands. The finest architects and artisans from all around Europe were recruited to transform St. Petersburg into a treasure trove of Baroque palaces and churches. ⚓

P*ort appeal rating*

Adventure	★★★★
Entertainment	★★★½
Romance	★★★½
Cuisine	★★½
Shopping	★★½

The Hermitage is a treasure trove for lovers of fine art and architecture.

St. Petersburg's Pushkin Palace, a breathtaking stop for visitors, was modeled after France's Versailles.

Left: St. Petersburg's architecture reflects an eclectic blend of Asian and European influences.

Classical music echoes through the Pushkin Palace, perfectly complementing the atmosphere of royal pomp.

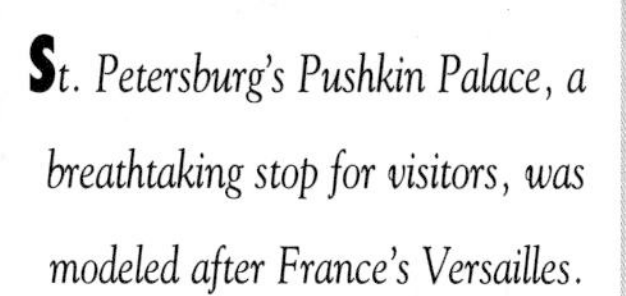

Golden domes grace the St. Petersburg skyline.

Overleaf: The walled city of Dubrovnik once contested Venice for dominance of the Adriatic, and remains a magical destination.

THE VITALITY OF

DUBROVNIK

"Liberty is worth more than gold," is one of the ancient mottoes of this stunning Adriatic city, which has managed to retain its distinctive flair through more than a thousand years of tumultuous Balkan history. It's nothing short of a miracle that Dubrovnik survived the recent disintegration of Yugoslavia with only minor damage. This southernmost Croatian city remains one of Europe's best-preserved medieval towns. Broad stone ramparts surround an old town suffused with cobblestone streets, historic monasteries, and astonishing palaces. ⚓

Port appeal rating

Adventure	★★★★
Entertainment	★★½
Romance	★★★★
Cuisine	★★★
Shopping	★★½

CONTRIBUTORS

CHRISTOPHER P. BAKER

A member of the Society of American Travel Writers and National Writers Union, Christopher P. Baker is the recipient of numerous awards for excellence in writing, among them a Benjamin Franklin Award for *Best Travel Guidebook, 1995*; the Lowell Thomas Travel Journalism Awards for Best Environmental Tourism Article, 1996; Best Guidebook, 1995; Best Travel News Investigative Reporting, 1993; and Best Self-illustrated Article, 1992. Baker has been interviewed on national radio and TV shows, and is a frequent guest speaker at trade shows and other events. He has been a consultant for INTEL Corporation and is also a faculty member of the lecturing staff of Cunard Line Limited. Baker lectures frequently aboard cruise ships.

Baker is an acknowledged expert on Cuba and has written about it for publications as varied as *World & I*, the *National Times*, the Organization of American States's *Americas*, the *New York Daily News*, *Esquire*, *Discovery*, and *Travel Agent*. He has escorted group tours to New Zealand, Hong Kong, Korea, Cuba, and England.

ROBERT W. BONE

Robert W. Bone is a veteran newspaperman and magazine and book editor, who has been writing on travel subjects since he worked in Europe with the late guidebook guru, Temple Fielding, 1968-1971.

From his Oceanic home base in Hawaii, Bone now travels the globe, specializing in cruises. He has also written four travel books, and his articles appear in the travel sections of the *Chicago Tribune* and several other major newspapers throughout the U.S. and Canada. Bone is also a photographer whose images have appeared in magazines and articles about travel.

JAY CLARKE

Jay Clarke, longtime travel editor of *The Miami Herald*, went on his first trip abroad at the age of five and has been traveling ever since. He has visited all fifty states, traveled to more than one hundred countries, and sailed on more than four dozen cruise ships. His articles have appeared in every major newspaper in the United States as well as a number of magazines and foreign publications.

Among the honors he has received are awards from the Society of American Travel Writers Foundation, the Pacific Area Travel Association, the Society of Newspaper Design, and honors from several states and foreign governments. He has been an officer and director of the Society of American Travel Writers and currently is a trustee of the SATW Foundation.

WILLIAM FRIAR

William Friar grew up in the Panama Canal Zone and is the author of *Portrait of the Panama Canal* (Graphic Arts Center Publishing). He has also lived in India and Denmark and traveled to more than forty other countries on five continents.

These days he works as a features writer, columnist, and music critic for Knight Ridder Newspapers. His work has appeared in *The Miami Herald*, *Arizona Republic*, *San Jose Mercury News*, *Orange County Register*, *Oakland Tribune*, *Houston Chronicle*, and other publications. A graduate of Stanford University and the Columbia University Graduate School of Journalism, he lives in San Francisco. He can be reached at wfriar@yahoo.com.

GEORGIA I. HESSE

Author, photographer, and past travel editor for the *San Francisco Sunday Examiner*, Georgia Hesse contributes to numerous magazines and newspapers and has authored or coauthored several books on travel. A Fulbright Scholar from l'Universite de Strasbourg, France, Hesse has traveled to over 181 countries and is the recipient of numerous awards of excellence for her writing and her extraordinary accomplishments in the travel writing field. Hesse has received the Ordre du Merité from the French government; Chevalier l'Ordre de la Republique, Tunisia; and the Melva C. Pederson Award from the American Society of Travel Agents. She is the first recipient of the new Georgia Hesse Award for Travel Writing, Redwood Empire Association.

LORRY HEVERLY

Lorry Heverly is an adventure travel writer who has wandered through over one hundred countries. Her journeys have been documented in a variety of travel magazines: *International Travel News*, *Marco Polo Magazine*, *Specialty Travel Index*, *Travel World News*, *Modern Bride*, plus newspapers and web e-zines.

Heverly is also a road warrior reporter for CBS affiliate WAFB/TV9, Baton Rouge, and files travel reports from worldwide locations. Her global adventures have taken her to emerging tourism destinations like Ethiopia, Croatia, and Midway Island; on safaris in remote regions of Africa and the hillstations of the mystical Himalayas; and on visits to the backwaters of the mighty Amazon and through deserts, rain forests, and hinterlands around the world. Her underwater adventures have brought her face-to-face with sharks, manatees, dolphins, and manta rays in habitats of coral atolls and ancient shipwrecks.

BARB AND RON KROLL

Barb and Ron Kroll specialize in the fields of travel, photography, food, and drink. Together as a writing and photography team, they have logged over three million kilometers around the world in search of good travel stories. To capture the essence of the destination, the Krolls insist on firsthand experience, be it dipping into an icy river after a Finnish sauna, feeding the sharks underwater in Bora Bora, sleeping next to human skulls in an ex-headhunter's longhouse in Borneo, fishing for piranha in the Amazon, or jumping out of a plane for a story on sky-diving. On one memorable two-year trip that brought them one hundred thousand kilometers by van through Europe, they skied with a member of the French national ski team and acted in a passion play (the first foreigners to do so in its centuries-old history).

Ranging the world from Anchorage to Zurich, the Krolls have ridden elephants in India and camels in the Tunisian Sahara, slept in yurt tents in Inner Mongolia, danced with Tuareg tribesmen in Africa, crossed the Atlantic in a Russian ship through two hurricanes and a gale, and were the first journalists to visit the Tehran bazaar after the Iranian revolution. Their stories and photographs are published throughout North America as well as internationally in a myriad of newspapers and magazines.

HARVEY LLOYD

Harvey Lloyd is an internationally renowned aerial and adventure travel photojournalist, writer, director, and poet based in New York City. He has traveled over a million and a half miles on worldwide photography assignments. Lloyd is the recipient of numerous awards for his aerial and adventure travel photography, picture books, documentary films, and multiscreen shows including the Pacific Area Travel Association's 1994 Gold medal for photography of China; The Art Director's Club Gold and Silver medals for the Royal Viking Line advertising campaign and the Club's first award ever for multimedia productions; a Cine Golden Eagle; a Mercurio d'Oro from the Venice Industrial Film Festival; the Grand Award—Photo Essays from the German magazine *Bildeerzeit*; and a Silver Anvil from the Public Relations Society of America for photojournalism on volunteer elder teachers.

Active in the field of eco-tourism, an ardent environmentalist and student of earth sciences and anthropology, Lloyd is an associate editor of *Ocean Realm Magazine* and a contributing editor (photography) to *Porthole Magazine*. His picture book on how to photograph from the air is the first and only book on this art. In addition to his worldwide adventure travel photography and picture books, Lloyd creates, writes, and directs documentary motion pictures and multiscreen sound and light shows. Lloyds' essays, poems, and photography have been published in magazines all over the globe.

GARRY MARCHANT

After working on small newspapers in British Columbia, Garry Marchant traveled to Rio de Janeiro, where he became editor of the *Brazil Herald*, the country's only English-language newspaper. A freelance writer, he has been based in Hong Kong for the past nine years, contributing to international publications.

Marchant's work has appeared in anthologies *Away From Home, Canadian Writers in Exotic Places* (1985), *Our American Cousins*, (1987), *That Reminds Me . . . Canadian authors relive their most embarrassing moments* (1990), and *Traveler's Tales Hong Kong*, (1996). He has traveled to more than 230 countries, on every continent (including the Antarctic).

JOHN MAXTONE-GRAHAM

John Maxtone-Graham is a world-famous maritime historian, author of three classics about passengers on shipboard: *The Only Way to Cross*, *Liners to the Sun*, and *Crossing & Cruising*. His most recent book is *Titanic Survivor*, published by Sheridan House.

In addition to this writing, Maxtone-Graham is an articulate and compelling lecturer. He spends about a third of each year lecturing to cruise ship passengers all over the world and at home in his native Manhattan at the Metropolitan Museum of Art.

Maxtone-Graham is the son of a Scot father and American mother, so it is small wonder that he remains both preoccupied and transfixed by the ocean that divides both halves of his life. From the garden of his Manhattan brownstone, Maxtone-Graham and his wife Mary can hear the whistles of ships sailing from the North River Passenger Ship Terminal five minutes away.

JO BETH MCDANIEL

Jo Beth McDaniel is a Southern California writer whose work has appeared in many international publications, including *Life*, *Islands*, *Reader's Digest*, and the *Travelers' Tales* series of books.

WILLIAM MILLER

William Miller is considered an international authority on ocean liners, both the great ships of the past as well as the current cruise vessels. He has written some forty-five books on the subject—from early liners to others in wartime, one on their fabulous interiors, and histories of such celebrated ships as the SS *United States* and the *Rotterdam*. He has also written over 750 articles on the great ocean liners and cruising and has made over two hundred voyages—from transatlantic and transpacific crossings to coastal runs on mail streamers to long voyages on tropic banana boats. Miller is also a guest lecturer aboard over fifty different cruise ships, and as a published marine photographer has appeared in many commercial videos, including one on the *Queen Elizabeth 2*, another on the artworks of the original *Rotterdam*, and in several television series—including A & E's *Floating Palaces* and The Learning Channel's *Castles of the Sea*. He also publishes his own quarterly about the passenger ship industry.

Miller has been historian at the Museum of the American Merchant Marine, chairman of the New York Branch of the World Ship Society, a board member of the Ocean Liner Museum, advisor to the Oceanic Navigation Research Organization, and former deputy director of the annual New York Harbor Festival. Also a contributing editor to the monthly *Ocean & Cruise News*, Miller is the recipient of numerous awards.

KATE SEKULES

Kate Sekules is an author and a veteran editor for *Food & Wine* magazine. She has traveled the globe writing for *New Yorker*, *New York*, *Vogue*, *Travel & Leisure*, *Travel Holiday*, *Bride's*, *Conde Nast Sports & Fitness*, *Savuer*, *Ritz Carlton* magazine, and *Harper's Bazaar*. Author of many guide books, including *Fodor's London*, *Frommer's Irreverent Guide*, and *By Night, London*, *CityPack New York*, her newest book, *Boxing For Girls*, will be out in 1999.

SHIRLEY STRESHINSKY

Shirley Streshinsky is a journalist, novelist, and travel essayist whose work has been published in some of the country's leading publications for the past twenty years. Her magazine articles have appeared in *Glamour*, *Redbook*, *Ladies Home Journal*, *Los Angeles Times Magazine*, *Preservation Magazine*.

Streshinsky is the author of seven published books: three works of nonfiction and four novels. Her travel essays have appeared in the *San Francisco Examiner's* Sunday travel section, in *San Francisco Focus*, in *Travel & Leisure* and *Conde Nast Traveler*, as well as many other publications in this country and abroad. She has been traveling to, and writing about, Hawaii for more than twenty years. She teaches a travel writing class at the University of California Extension at Berkeley.

JOSEPH R. YOGERST

Joe Yogerst is a well-known writer and editor based in San Diego, California. A world traveler, he has lived in Europe, Singapore, and the Caribbean. For three years, Yogerst was the managing editor of the award winning *Discovery* magazine in Hong Kong and during that time he received a Lowell Thomas Award from the Society of American Travel Writers. Yogerst has been a contributing editor to numerous publications including *Conde Nast Traveler* magazine in New York, and the regional editor of *Pacific Traveller* magazine in Hong Kong. His work has appeared in *The Washington Post*, *The Los Angeles Times*, *The San Francisco Examiner*, the London *Mail on Sunday*, *The Geographical Magazine*, the *International Herald Tribune*, and *Time*. He is presently working on a book for National Geographic.

Courtesy of Carnival Cruise Lines 282
Courtesy of Costa Cruise Lines 12-13, 29, 38-39, 47, 282
Michael Dakota 51
Rick Diaz 51, 53, 54
Mark Downey 194, 195
Lee Foster 21, 132-133, 142-143, 143
Kevin Garrett 68
Bob Grieser 92-93, 95, 140, 141, 131
Courtesy of Holland America Line 32-33, 34, 48-49, 53, 154, 283
Wolfgang Kaehler 206
Barb and Ron Kroll 64, 65, 212, 258, 260
Gary Kufner 55
Helga Lade/Peter Arnold Inc. 207
Danny Lehman 14-15
John and Lisa Merrill/Delimont, Herbig, & Assoc. 212-213
Bill Miller Collection 16, 40, 44, 45, 46
Andy Newman 50
Paul Souders/Delimont, Herbig, & Assoc. 166-167
Nik Wheeler 208-209, 254
Joyce Gregory Wyels 94

PARTICIPATING TRAVEL WRITERS

The editors of Voyages: The Romance of Cruising *wish to thank the following travel writers for their participation in rating the world's 100 most exciting ports of call.*

Mary Lu Abbott, consulting editor, Vacation Publications
Ludmilla Alexander, travel editor, *South Bay Accent Magazine*
Eric Anderson, travel editor, *Physician's Money Digest*
Molly Arost Staub, travel writer, Boynton Beach, Florida
Geri Bain, travel editor, *Modern Brides Magazine*
Millie Ball, travel editor, *The Times-Picayune*
Len R. Barnes, editor, *Michigan Living Magazine*
Carol Barrington, travel writer, Montgomery, Texas
James Y. Bartlett, managing editor, *Caribbean Travel and Life*
Patricia Bell, travel editor, *Gourmet Magazine*
Arline Bleecker, travel writer, Millington, New Jersey
Larry Bleiberg, assistant travel editor, *The Dallas Morning News*
Ethel Blum, travel writer, Aventura, Florida
Alys Bohn, senior editor, *Recommend Magazine*
Marybeth Bond, editor, *Traveler's Tales*
Robert Bone, owner, *World Travel Network*
Al Borcover, travel writer, Evanston, Illinois
Laurie Borman, editor-in-chief, *Endless Vacation*
Lester and **Patricia Brooks**, travel writers, New Canaan, Connecticut
J.D. Brown, travel writer, Eugene, Oregon
Jay Brunhouse, author, *Traveling Europe's Trains*
Laszlo J. Buhasz, travel writer, *The Globe and Mail*
Michelle Burgess, travel writer, Huntington Beach, California
Anne Campbell, cruise critic, America Online
Yvette Cardozo and **Bill Hirsch**, travel writers, Issaquah, Washington
Richard Carroll, travel journalist, *Motor Home Magazine*
Jay Clarke, travel editor, *The Miami Herald*
Anne Z. Cooke, editor, The Syndicator Travel News Service
Patti Covello, writer, *Access Cruise*
Robert C. Cross, travel writer, *Chicago Tribune*
Georgina Cruz, travel writer, Miramar, Florida
Randy Curwen, travel editor, *Chicago Tribune*
Alison DaRosa, travel editor, *San Diego Union Tribune*
Guy Demarino, writer, *Edmonton Sun*
Sophia Dembling, travel writer, Dallas, Texas
George Devol, writer, *Ocean and Cruise News*
Jerry Flemmons, travel writer, Granbury, Texas
Lee Foster, travel writer/photographer, Foster Travel Publishing
Carol Fowler, travel editor, *Contra Costa Times*
Robin Fowler, managing editor, Vacation Publications
Larry and **Barbara Radin Fox**, travel writers, Kensington, Maryland
William Friar, reporter, Knight Ridder Newspapers
Robert Fried, writer/photographer, Robert Fried Photography
Barbara Gillam, travel editor, *Glamour Magazine*
Bill Gleasner, freelance photographer,
Diana Gleasner, travel writer,
Fran Golden, cruise editor, *Travel Weekly*
Arturo and **Maureen Gonzalez**, contributing editors, *Hurgman Magazine*
Peter Greenberg, travel editor, NBC *Today Show*
Judy Hammond, travel writer, Pacific Grove, California
Robert Haru Fisher, editor, Fisher Publications
Gloria Hayes Kremer, travel writer, Rydal, Pennsylvania
Mary Ann Hemphill, travel writer, Newport Beach, California
Brook Hill Snow, travel writer, Orlando, Florida
Vivian Holley, travel writer, Roswell, Georgia
Dave G. Houser, travel writer, Ruidoso, New Mexico
Carla Hunt, travel writer, New York, New York
Michael Iachetta, writer, Reed Travel Group/Star Service
Robert N. Jenkins, travel editor, *St. Petersburg Times*
Stephen Jermanok, travel writer, Newton, Massachusetts
Susan Kaye, travel writer, Aspen, Colorado
Evelyn Kieran, travel writer, San Anselmo, California
Sally Kilbridge, managing editor, *Brides' Magazine*
Helmet Koenig, travel writer, New York, New York
Katy Koontz, travel writer, Knoxville, Tennessee
Michael Lachetta, travel writer, Scarsdale, New York
Paul Lasley and **Elizabeth Harryman**, co-hosts / producers, *On Travel Radio*

Florence Lemkowitz, travel writer, St. Petersburg, Florida
Marcia Levin, travel writer, Hollywood, Florida
Christopher Lofting, travel writer, New York, New York
Gene and **Adele Malott**, editors, *The Mature Traveler*
Theresa N. Masek, cruise editor, *TravelAge West*
John Maxtone-Graham, historian and author, New York, New York
Maribeth Mellin, travel editor, *San Diego Magazine*
Phyllis Meras, former travel editor, *The Providence Journal*
Laurence Miller, travel writer, Miami, Florida
Jerry D. Morris, travel editor, *The Boston Globe*
Nancy Muenker, travel writer, Muenker Media
Bill Panoff, publisher, *Porthole*
Wendy Perrin, consumer news editor, *Conde Nast Traveler*
Percy Rowe, travel writer, Etobicoke, Ontario, Canada
Joy Schaleben Lewis, travel writer, Milwaukee, Wisconsin
Jill Schensul, leisure editor, *The Record*
Joe Scholnick, travel editor, Capitol News Service
Marshall Schwartzmann, travel writer, Rutherford, New Jersey
Joan Scobey, travel writer, New York, New York
Lynn Seldon, travel writer, Richmond, Virginia
Kay Showker, travel writer, New York, New York
Jean Simmons, travel columnist, *Dallas Morning News*
Joel Sleed, travel columnist, Newhouse Newspapers
Libby Smith, travel editor, *Arkansas Democrat-Gazette*
Alan Solomon, travel writer, *Chicago Tribune*
Phillip Sousa, travel writer, San Diego, California
Candyce H. Stapen, family travel specialist, *Cruise Vacations with Kids*
Janet Steinberg, travel editor, *The American Israelite*
Dorothy Storck, travel writer, Chicago, Illinois
Toni Stroud, consumer travel writer, *Chicago Tribune*
David Swanson, travel writer, Dorchester, Massachusetts
Mim Swartz, travel editor, *Rocky Mountain News*
Joan Tapper, editor, *Islands*
Gerry Tobin, travel editor, *San Juan Star*
Myra Waldo, writer, *Myra Waldo's Travel Guides*
Claire Walter, travel editor, *Skiing Magazine*
James Wamsley, contributor, *Self*
Douglas Ward, author, New Milton, New Hampshire
Catherine Watson, travel editor, *Minnepolis Star Tribune*
Deborah Williams, travel writer, Holland, New York
Jane Wooldridge, assistant features editor, *Miami Herald*
Joe Yogerst, travel writer, San Diego, California

WORLD'S LEADING CRUISE LINES

The World's Leading Cruise Lines are all about choosing a cruise vacation that fits your individual style. Each cruise line is a leader in a particular style of cruise vacationing, and together they provide you with a range of options so you can choose the cruise experience that's made to order for your vacation.

WORLD'S LEADING CRUISE LINES

With its signature "Fun Ship" cruise vacations, Carnival offers contemporary sparkle and value for couples, families, seniors, and groups. Whether sailing to the Caribbean, the Bahamas, the Mexican Riviera, Alaska, the Panama Canal, or Hawaii, you'll have the time of your life and receive the best vacation for your money.

1-800-CARNIVAL

www.carnival.com

COSTA CRUISES

Italian Style

Italian charm and exotic destinations make Costa Cruises a distinctive choice. Authentically Italian in its design, activities, and cuisine, Europe's number one cruise line offers sailings throughout the Caribbean, Mediterranean, Northern Europe, and South America with comfort and hospitality that is not just great, but *Magnifico*!

1-800-33-COSTA

www.costacruises.com

Cunard maintains the tradition of luxury that has marked its 158-year history. With elegant ambiance and distinctive onboard service, Cunard's ships serve nearly all regions of the world in the grand tradition of the classic ocean voyage. A Cunard cruise is the ultimate travel experience for those who expect the best.

1-800-7-CUNARD

www.cunardline.com

For more information, see your local professional travel agent.

INDEX